The International Politics of Space

T0188293

The year 2007 saw the fiftieth anniversary of the Space Age, which began with the launching of Sputnik by the Soviet Union in October 1957. Since that time, the development of space technology has revolutionised many aspects of life on Earth, from satellite television to mobile phones, the internet and micro-electronics. It has also helped to bring about a revolution in the use of military force by the most powerful states.

Space is crucial to the politics of the postmodern world. It has seen competition and cooperation in the past fifty years, and is in danger of becoming a battlefield in the next fifty. *The International Politics of Space* is the first book to bring these crucial themes together and provide a clear and vital picture of how politically important space has become, and what its exploitation might mean for all our futures.

Michael Sheehan analyses the space programmes of the United States, Russia, China, India and the European Space Agency, and explains how central space has become to issues of war and peace, international law, justice and international development, and cooperation between the world's leading states. He highlights the significance of China and India's commitment to space, and explains how the theories and concepts we use to describe and explain space are fundamental to the possibility of avoiding conflict in space in the future.

This landmark book will be of great interest to students of international relations, space politics and security studies.

Michael Sheehan is Professor of International Relations, University of Swansea. He has taught courses on the international politics of space for the past twenty years, and is the author of numerous books including: *International Security: An Analytical Survey*; *National and International Security*; *The Balance of Power: History and Theory* (also published by Routledge); *Arms Control: Theory and Practice*; and *The Arms Race*.

Series: Space Power and Politics
Series editors: Everett C. Dolman and John Sheldon
*Both School of Advanced Air and Space Studies, USAF Air,
Maxwell, USA*

The International Politics of Space

Michael Sheehan

Routledge
Taylor & Francis Group

LONDON AND NEW YORK

First published 2007
by Routledge
2 Park Square, Milton Park, Abingdon, Oxon OX14 4RN

Simultaneously published in the USA and Canada
by Routledge
270 Madison Ave, New York, NY 10016

*Routledge is an imprint of the Taylor & Francis Group, an informa
business*

British Library Cataloguing in Publication Data
A catalogue record for this book is available from the British
Library

Library of Congress Cataloging-in-Publication Data
Sheehan, Michael (Michael J.)
The international politics of space / Michael Sheehan.
    p. cm. – (Space power and politics)
  Includes bibliographical references and index.
  1. Outer space–Exploration–Political aspects. 2. International
  relations. 3. Outer space–Exploration–International
  cooperation. I. Title.
JZ3877.S54 2007
629.45–dc22   2007015930

ISBN10: 0–415–39807–X (hbk)
ISBN10: 0–415–39917–3 (pbk)
ISBN10: 0–203–93390–7 (ebk)

ISBN13: 978–0–415–39807–7 (hbk)
ISBN13: 978–0–415–39917–3 (pbk)
ISBN13: 978–0–203–93390–9 (ebk)

To my uncle, John O'Dea

Contents

Introduction

This is a book about outer space. Unlike most books about space, however, it is not primarily concerned with technology, at least not directly. Rather, it is focussed on the *politics* of space. Technology is important in this account certainly, but not for its own sake, or even essentially for the missions of exploration or military tasks that it makes achievable. Rather it is important to the extent that its achievements, and the resources devoted to bringing them about, say something important about the political motivations and expectations of the governments that committed themselves to those programmes.

The vast majority of the books that have been published dealing with space policy are examinations of the military uses of space, and particularly the question of whether weapons should, or should not, be deployed there. These are important issues, but the overwhelming focus on them is a barrier to an understanding of the broader political dimensions of the use of space. Politics has always been at the heart of mankind's exploration and utilisation of space, and the space programmes themselves have never been able to transcend terrestrial international politics; they have only reflected it. As Walter McDougall put it, 'despite the flights of fancy of some space law theorists, there was no "escape velocity" that took one beyond the political rivalries of this world'.[1]

A study of the international politics of space therefore provides both a corrective to the idea that space programmes are science-driven bureaucracies somehow aloof from the harsher realities of politics, and reveals case-studies of themes that are familiar in other dimensions of international relations. In space, as on Earth, we see the political power of ideology and nationalism, the use of propaganda and foreign aid, the centrality of questions of 'national security' and the pursuit of that security through the acquisition of military capabilities, tensions between the richer, more industrially advanced states, and the poorer countries of the 'South', efforts to use the integration of national policies to further the unity of Europe, the evolution of understandings of security to embrace social, environmental and economic dimensions and so on. There are few, if any, features of contemporary global politics that do not have their echoes in the utilisation of space.

Nor should the military significance of the exploitation of space be underestimated. Clausewitz famously declared that war is 'a continuation of politics with an admixture of other means'. So too is the exploration and utilisation of outer space. There have been times in the past 50 years when public perceptions of space have seemed to contrast a pristine idealism of space exploration represented by agencies like NASA and ESA, with the sordid programmes of the armed forces, determined to sully the celestial realm with their efforts at the 'militarisation' of space.

This is a misleading perspective. Space has always been militarised. Military considerations were at the heart of the original efforts to enter space and have remained so to the present day. Efforts to turn space into an entirely non-militarised 'sanctuary' may be commendable, but if they were achieved, they would not be a successful defence of space from the looming threat posed by militarisation. Rather they would represent a dramatic reversal of policy, the recreation of the human realm of outer space in a form that has never in fact existed since the dawn of the space age.

Space and politics are, and always have been, inseparably interlinked. The central driving force for all space programmes has been political objectives. Space programmes have reflected and implemented the prevailing national and international ideologies of the time, whether they be power politics, communist internationalism, European integration, national self-determination or anything else. Space policy cannot be divorced from politics and never has been. In 1972, as the United States prepared to send Apollo 17, the last manned mission to the Moon, the Black September group, who were responsible for the Munich Olympics massacre in the same year, threatened to sabotage the launch.[2]

But while space programmes have been shaped by the politics of the past half-century, in turn the utilisation of space has helped to shape the politics of the post-modern world, providing iconic images of planet Earth to energise the environmental and peace movements, stabilising the Cold War through deterrence and arms control, and thereby helping to avoid a nuclear Armageddon that threatened humanity for nearly half a century, producing satellite communications systems that gave some tangible meaning to the fuzzy concept of globalisation, while doing so in a way that continued to distinguish sharply between the reality enjoyed, or even imagined, by humanity's 'haves' and 'have-nots'.

This is the Space Age. Politicians have struggled unsuccessfully to find a term that might capture the meaning of the 'post-Cold War' world as anything more than a post-script to the era that preceded it. But in the *longue duree* of historical perspective, it is likely that both the Cold War and the 'War on Terror' will come to be seen as no more than dramatic historical episodes. It is the fact that it was in the second half of the twentieth century that humanity first moved into, and began to exploit, the potential of space that is most likely to define our age in the decades and centuries to come, as the voyages of Christopher Columbus and Vasco da Gama have come to define the early modern period, for whatever political requirements originally motivated it, future generations 'will surely consider the exploration of the Solar System to be one of the most outstanding achievements of humankind'.[3] The advent of spaceflight produced a fundamental historical disjuncture, only dimly perceived at first, between industrial modernity, and the post-modernity of the information age. The information age, in all its manifestations, has in turn brought into being, for the first time in human history, a truly planetary international political system. The space age is the age of global politics.

With the space age humanity has achieved unprecedented power, but has also come to experience, and to be fully aware of, unprecedented vulnerability. The ability to be simultaneously aware of both is the result of the unprecedented wealth of information and alternative ways of interpreting it, that space exploration and exploitation brought about through satellite technology and the computing revolution.

The half-century of spaceflight has brought with it a certain degree of complacency about what has been achieved, as what once seemed fantastic very quickly came to seem banal. Space, as the military are fond of pointing out, is the 'new high ground', and the high ground has always been sought in war for the military advantages it brings with it. But by analogy, we speak also of seizing the 'moral high ground', and there is an important sense in which outer space still has the potential to be either. The space programmes allow us to stand on the shoulders of giants, and gain a perspective on global politics that is difficult to achieve from ground level, or ground zero.

In order to begin to gain such a perspective, it is first of all necessary to see where we are, and where we have come from. There are different ways in which we can think about space, and different theoretical paradigms that we can use to interpret its meaning and question its use. To engage fully in contemporary debates on the 'weaponisation' of space, for example, it is important to understand what purpose space has, and has not served to this point in history, because the contemporary military space programmes

> are haunted most immediately by the prospects for greater destructive capacity that they portend, but also by alternative visions for the use of space that they preclude. Marcuse argues that 'naming the things that are absent, is breaking the spell of the things that are'.[4]

Focusing on the importance of military space always runs the risk of losing sight of other parts of the space context, particularly the dynamism of cooperative international space activity, and this subject is therefore looked at broadly in Chapter 4 and specifically in relation to European integration in

Chapter 5. It is also a theme returned to in the discussion of the international space station in Chapter 11.

The significance of the American and Soviet/Russian programmes is a recurrent theme throughout the book, and the shaping power of the original superpower space race cannot be overestimated. The heavy emphasis on this period in the initial three chapters reflects the fact that it was during these years that a number of subsequently crucial political themes were first sharply revealed, and these remain central to an understanding of space politics in subsequent decades. The military dimension of this rivalry was also fundamental and the military uses of space are clearly vital to any understanding of the political importance of space. It is important to analyse this, not only in terms of the immediate support functions that space systems now offer to the military, both at the strategic and tactical levels, but also to examine the broader strategic and political context in which space power concepts relate to both military and civil uses of space. This is done in Chapters 6 and 7. In recent decades, the dominance of space policy by the developed world and the state has evolved as important actors have emerged in the developing world and outside national governments. China and India have become leading space powers, fuelling their own development strategies, and in the case of China, acting as a spur to the re-energising of the space programmes of developed countries such as the United States and Japan, stung by the Chinese challenge. Chapters 9 and 10 examine the nature and significance of the Indian and Chinese space programmes.

The purpose of this book is to encourage a broad perspective on the politics of space. It has been necessary to sacrifice some depth in order to acquire this breadth. Not every interesting aspect of contemporary space policy is covered. There is little specific attention, for example, to the vigorous commercial space industry, both in terms of launchers and applications satellites, that is separate from the national space programmes that dominated the first decades of the space age. The emergence of this sector and the developing 'space tourism' industry are reflections of the evolving nature of the state produced by the continuing transition from modernity to post-modernity, and as understandings of politics and the state change, this will be reflected in the understanding and utilisation of space. Similarly, not every significant national space programme is examined, the examples selected were chosen because they have something specific to say about the political uses of space.

Chapter I

Perceptions of space and international political theory

Space is a great emptiness into which we often project our dreams.[1]

Introduction

How should we think about space? It makes a difference how we do, because although we as humans live in a physical universe, much of the 'world' we inhabit is intersubjectively constructed through our mutual understandings of what constitutes reality. We act in terms of our beliefs, values, theories and understandings of the 'reality' we perceive. It is also important to remember that the way in which such a consensus on understandings of reality is constructed is not an entirely innocent exercise. As Cox pointed out in relation to the production of theory, 'theory is always for someone and for some purpose'.[2] By firmly establishing a specific perception of outer space, a dominant narrative helps to shape a particular reality. We perceive outer space in a particular way, as a particular kind of realm, in which certain types of activity are possible, even expected, while others are frowned upon or specifically forbidden.

When there are alternative conceptions available, a particular visualisation is likely to favour the interests of some states more than others. In 1957 space was essentially a *tabula rasa*, a blank page on which humanity was free to write whatever it chose. But it brought with it pre-existing values and behaviour patterns. The major powers who first entered outer space had policies and belief systems structured by the 'lessons' of previous decades, and particularly by the catastrophe of the Second World War and the bitter peace that came to be called the Cold War. In the decades that have followed, policy makers, scientists and advocates of space exploration have contested opposing understandings of the meaning and purpose of outer space for humanity. The image we have of the extra-terrestrial realm ought to be such a contested terrain, for what we perceive space to be shapes our views of how it should be exploited, and this has very real implications for political, economic and environmental development on Earth.

It was only with the advent of the first satellite that space became an ontological reality directly experienced by mankind. But even prior to that point it had never been truly a vacuum in terms of the way that it was perceived by humanity. Space was both an environment in which many possibilities could be imagined and a fruitful source of metaphorical meanings, such as freedom, opportunity and infinite possibilities, and its multitude of possible interpretations included those that were ambiguous or incompatible.[3] For millennia humans had speculated about the nature of what lay beyond their world, and had habitually placed the realm of the gods that they worshipped in the dimension that lay out of sight above their heads. The night skies were a place of beauty and mystery, and these cultural understandings of space have played a part in maintaining resistance to certain developments in the use of space, most notably the extension of terrestrial weapons and warfare beyond Earth's atmosphere. Such a development can be seen as threatening what the Dutch call *vergankelijkheid*, the transitory nature of what is beautiful and magnificent.[4] The desire to maintain space as a war-free sanctuary certainly existed immediately prior to the beginning of the space age. As early as 1952 the International Congress on Astronautics voted to ban its members from using astronautical research for military purposes.[5]

The idea of preserving certain geographical areas as demilitarised sanctuaries has a historical pedigree as old as the space age itself. An entire class of arms control agreements, the 'non-armament treaties' have been concluded over the past five decades, designed to 'prevent military competition from being introduced into an area that had hitherto been free of such activity'.[6] This group of treaties includes the 1967 Outer Space Treaty, but the first such agreement had come as early as 1959, with the Antarctic Treaty. All the agreements were based on the belief that it was both desirable and possible to maintain certain parts of the human environment as sanctuaries in terms of military activity.

Advocates of space militarisation have been very aware of the power of these conceptualisations. One such advocate noted that the idea of preserving space as a *sanctuary* from militarisation is commonly held, and that 'in using the term "sanctuary", critics of the military use of space mean not only a place of refuge or asylum, but a sacred and holy place secure from the baser instincts of men. No wonder military programs such as Star Wars or antisatellite (ASAT) warfare have elicited such a strong reaction'.[7] Even President Eisenhower's science adviser felt that the launch of Sputnik I by the Soviet Union in 1957 had stirred 'atavistic, subtle emotions about cosmic mysteries' and 'an instinctive, human response to astronomical phenomena that transcend man's natural ken'.[8]

The ambiguity and incompatibility of differing interpretations however lent itself to political exploitation. While the United States agreed with other states that space belonged 'to all mankind' for example, what it understood by this phase was no more than the celestial equivalent of the idea of the 'freedom of the seas',[9] while other countries invested far greater philosophical and political meaning into the concept – and assumed that the United States did the same. The ambiguity and conflicting motivations that have historically surrounded space exploration mean that it is not always the case that there is a single or simple explanation for particular space programmes or missions.

In discussions of space policy, idealism and realism continue to clash. The debates remain potent, because apart from the brief American *Apollo* expeditions to the Moon between 1969 and 1972, human beings have remained locked in low-Earth orbit even a half century after the beginning of the space age. To date only 24 human beings have ever viewed their planet from the deep space beyond Earth orbit, and the difference in perspective between Earth orbit and deep space is tremendous. 'The orbital astronaut experiences the planet as huge and majestic, while from afar it is tiny, beautiful and shockingly alone'.[10] It is obvious that to date the history of space exploration and utilisation suggests a conclusion that 'the Space Age would neither abolish nor magnify human conflict, but only extend politics-as-usual to a new realm'.[11] Nevertheless, so long as almost all of the Solar System and beyond remains essentially virgin territory, advocates of the 'sanctuary' perspective can argue that all is not lost, despite the unpromising historical track record of humanity in conquering new territories in the quest for knowledge.[12]

In international relations theory, it was customary during the late Cold War period to speak of an 'inter-paradigm debate', between contending world views of international relations, realism, liberalism and Marxism. The concept is somewhat misleading, in that little or no genuine 'debate' has occurred between the proponents of the different 'paradigms', other than perhaps between the squabbling siblings of realism and liberalism; nevertheless the idea of clearly distinct theories is analytically useful in allowing comparison of different perspectives and relating them to differing policy implications. It is worthwhile therefore to consider how the different paradigms of international relations might influence our understanding and interpretation of space, and also to consider what are the paradigms within which space itself has been considered in the past half-century.

Looked at from a chronological historical perspective there is some logic to addressing the claims of realism first. The Space Age began with the launch of Sputnik I in 1957, at a time when the Cold War was at its height and *realism* exercised a hegemonic dominance within the academic discipline of international relations. Studies of the early 'space race' between the superpowers typically employ key realist themes to explain the space competition. This is made easier by the fact that in this period it was classical realism that was dominant, exemplified by the writings of Hans Morgenthau and Raymond Aron, and this version of realism was richer and more nuanced than the narrow neorealism characteristic of the 1980s and thereafter. Realists like Morgenthau saw politics as 'a struggle for power and unilateral advantage',[13] in which the operation of the balance of power was central. In the absence of world government an 'international anarchy' exists and this self-help system produces a 'security dilemma' in which attempts by states to increase their own security lead to increasing insecurity because each state views its own defence efforts as legitimate and non-threatening, but those of other states as unnecessary and hostile.[14]

From the classical realist perspective the space race is explained by the competition for power between the superpowers, but the 'power' in question is a multifaceted amalgam of different forces ranging from tangible military capability to unquantifiable degrees of prestige. A space programme could contribute to overall power by confirming or suggesting capabilities in a range of other areas, such as long-range missiles and technological expertise. In the classical-realist approach domestic political explanations are also significant in a way that they are not in neorealism and therefore the internal political dynamics of the American and Soviet political systems are also an important part of the equation.

There are aspects of the history of national space programmes that seem to validate elements of both classical realist and neorealist international relations theory. The early superpower space programmes clearly lend themselves to a *realist* interpretation. The USA and USSR saw themselves as acting within an international anarchy in which the security dilemma was particularly dramatic as the implications of mutual nuclear capability sank in. 'National security', defined as military security from the armed forces of the opposing superpower, became the unquestioned priority of both states' leaderships. The relationship between the USA and USSR was understood in both countries as being competitive at best, conflictual at worst. The motivating driver of both programmes was the acquisition of military capability, both in terms of missiles able to deliver nuclear weapons, and satellites capable of securely performing reconnaissance missions over the adversaries' territory. While public attention focussed on the unmanned and manned competitive space programmes, these too were simply part of the global competition for international leadership in an era when direct military confrontation was increasingly unthinkable. The civilian and military programmes were linked to the extent that the former diverted attention from the latter, and in some cases, such as the US Explorer/Corona satellite, was used as a deliberate cover for military activities.

For realism, states living in the fearful world produced by the international anarchy will seek 'opportunities to shift the balance of power in their favour. At the very least, states want to ensure that no other state gains power at their expense'.[15] What was critical about the 1950s was that the superpowers' possession of nuclear weapons led them to believe that maintenance of the balance of power by traditional warfighting methods might prove suicidal, and therefore alternative conceptions of the balance and the way that it might be manipulated became crucial. In this respect, a superpower competition in space launching was an attractive alternative to a nuclear conflict in terms of demonstrating relative power capabilities.

The movement into space was very much an outgrowth of the terrestrial superpower competition for planetary hegemony and their respective space capabilities grew out of the strategic nuclear arms competition.[16] Nevertheless, as a theatre of political interaction, the space environment was responsive to changes in the world system, as was reflected in the emergence of Europe, China, India and others as players in the drama. Again, realism had no difficulty in explaining this in terms of the gradual evolution of the international system from a fairly rigid bipolarity in the immediate aftermath of the Second World War, to a more complex multipolarity as the 1960s gave way to the 1970s.

The vacuum of space did not remain a vacuum in political terms once the Soviet Union issued the ideological challenge represented by the launching of Sputnik in 1957. On the contrary, it soon became an emblematic example of the same power politics that characterised relations between the major powers on Earth. In addition, the movement into space brought about a new criterion for determining the gradation of power and the allocation of prestige in the global community. As Knorr noted early in the space era, 'if space activities are conducted on a large scale and on top (actually, as part) of the arms race, only a few individual countries will be able to muster the resources for effective participation'.[17]

This was an early example of the kind of thinking that in the late Cold War and post-Cold War era would be called 'techno-nationalism'. In this view, the economic and political power associated with access to the most advanced technology has made it the crucial determinant of international power and status. In the contemporary international system the development of advanced technology has now become the key system variable in the way that military power and alliance membership previously was, and geo-technological manoeuvring has replaced geopolitical rivalry in the global competition for status and political influence.[18]

Technonationalism at first sight seems remote from the traditional realist preoccupation with military capabilities as the ultimate measures of power within the international system. However, it is certainly possible to include it within the broader measurement of power associated with classical forms of realism. In addition, while it is distinct from a purely military measurement of power, there is nevertheless a significant overlap between the kinds of technologies typically advanced by technonationalist regimes and contemporary indicators of military prowess. The space programmes of India, China and Europe, for example, have delivered enhanced military capabilities through the development of long-range launchers and satellite systems with a wide variety of military utility, including reconnaissance, communications, navigation and meteorology. The complex interrelationship between technonationalism as such, and enhanced military capability, helps to explain why developing countries such as China and India chose to invest in expensive space programmes in the face of the enormous domestic poverty and underdevelopment with which they were struggling. The realist emphasis on the importance of reputation and prestige explains why such countries would eventually choose to move into the hugely expensive commitment of a manned spaceflight programme, as both China and India are now doing. Techno-nationalism is therefore a useful concept within a frame of reference that focuses on the motivations of emerging great powers such as China.[19]

The 'space race' was a dramatic competition for influence and prestige, conducted by two societies whose different ideologies both stressed the crucial historical importance of competition. For powerful states possessing an abundance of technological, scientific, financial and political capability it was easy to argue that *competition* is beneficial, because it produces benefits for all.[20] However, for those states and peoples concerned with the human benefits that could come from the practical application of space activity (especially in economic, environmental and developmental terms) there has been a contrasting emphasis on *cooperation*. Thus, space activity brought an alteration in the visible measurement of power, in its image, but not in the underlying fundamentals. From its inception, space activity brought a change in the image of power to one of scientific and technological progress, but the underlying relationship of 'haves' and 'have-nots', of those who possess power, and those over whom it is exercised, was not affected. Thus authors of international relations texts could confidently assert that space technology had come to symbolise the power of the Soviet Union and the United States.[21]

Certainly, there is no doubt that considerations of power and prestige were at the heart of the early superpower space programmes and they have remained central to the motivations of the actors in the more pluralistic space environment that has emerged in the past quarter of a century. Taking issue with those who argued that superpower space cooperation might help bring about a thaw in the Cold War, a 1967 study by Cash stated firmly that 'the prospect of space cooperation providing an instrument for changing reality was rejected as preposterous by those constructing the program; it was never considered. Given the perception of reality that dominates the present world one could hardly expect otherwise'.[22]

While classical realism saw state power as embracing more than the military dimension, nevertheless military power was emphasised to a greater extent than other forms. Apart from its specific uses in deterring attack, supporting allies, acquiring resources and so on, it was seen as a shield behind which all other tools of influence could be exercised. The military potential of space was recognised at the outset of the space age.

The movement into space opened up 'unprecedented possibilities of a military nature'.[23] Even before the end of the Cold War the growing importance of satellites in the conduct of strategy and foreign policy encouraged the belief that 'conventional wisdom regarding the conduct of world affairs is rapidly being dispelled'.[24] The strategic significance of space has been brought about by the tremendous advances in communications, surveillance and navigation that they have made possible. Satellite systems have advanced to the point where they have become the eyes, ears and voices of the major powers. But, as the increasing dependency of great power strategy has become apparent, the satellites that make it possible have themselves become attractive military targets.

As with the skies in the early twentieth century, space evolved from being seen simply as an environment in which the use of force on the ground might be aided, to a dimension in which combat would take place, as each side sought to exploit the military use of space, and deny its use to the enemy. The logic of the *inevitability* of such developments is in line with the realist approach to international relations, and it is similarly a self-fulfilling prophecy to the extent that states act as if it was true.

Neorealism can also be felt to be validated by the convergence in goals that has occurred over the same period. By the mid-1980s the various space programmes had obvious similarities, but also important differences. A key feature of the neorealist explanation of international relations is the argument that the security dilemma compels states to behave in essentially similar ways if they are to survive and prosper. The constraints of the system drive states to become functionally alike in the security realm. There is evidence to support this claim in the evolution of several space programmes in the past three decades. The programmes of Japan and the European Space Agency, for example, originally had no military dimension, while those of China and India lacked a manned presence in space, nor did any of these national and international programmes seem to feel that these absences constituted a significant weakness. In the past two decades, however, the various programmes have become increasingly similar in terms of their content and objectives.

Europe and Japan have now added a military dimension, while China has acquired a manned programme and India has announced its intention to do so. These developments appear to validate the neorealist argument that states in the international system differ in capability, but exhibit a similarity in objectives and process, and indeed are obliged to do so by the nature of the system.[25] Neorealists like Waltz argue that states are obliged to be functionally alike, that they tend to operate with a similar range of instruments and to use them in remarkably similar ways, constrained only by the comparative resources available to them.

Against this, realist assumptions about the likelihood of competition in the international anarchy are not necessarily borne out by the history of space policy. For realists, states are not inclined to cooperate unless there are compelling reasons to do so, because of the mutual insecurity they experience under the security dilemma. Weber, for example, argues that international cooperation is likely to be limited, and where it does occur, will be 'tenuous, unstable and limited to issues of peripheral importance'.[26] In space policy, however, states have frequently sought out opportunities to cooperate and have often self-consciously seen this as a possible way to mitigate the dangers inherent in an adversarial relationship such as that between the superpowers during the Cold War,[27] or between China and Russia.

Some realist proponents allow for such cooperation. Glaser, for example ,argues that there will be circumstances where a state's best security strategy will be cooperation rather than competition.[28] For realists, statesmanship is about 'mitigating and managing, not eliminating conflict; seeking a less dangerous world, rather than a safe, just or peaceful one'.[29] There is clearly an appropriate place for international cooperation in such a world view, though it is not seen as overcoming the essentially conflictual nature of international relations. Thus, space activity brought an alteration in the visible measurement of power, in its image, but not in the underlying fundamentals. Given the dominance of realist thinking in the early years of the space age therefore, it was always likely that competition, rather than cooperation, would be the dominant political theme.[30]

Other theoretical approaches to the study of international relations can also be usefully applied to the politics of space since 1957. Liberalism and neoliberalism also lend themselves to the analysis of key aspects of space policy in the past half century, both in terms of the way that liberalism focuses on the domestic factors explaining national space policies, and in terms of neoliberalism's focus on the effects of international organisations and international regimes. This approach has obvious utility for the study of the development of the international law of space, the role of international institutions such as the European Space Agency and the United Nations Organization, and the practice and significance of international cooperation in the space field.

Liberalism provides a mechanism for seeing national space policy not as the output of a unitary national government, but as the end product of complex political interactions between domestic actors. United States space policy from this perspective is not simply something produced by NASA, but rather the result of political bargaining processes between NASA, the Department of Defense, the State Department, Congress, US state governments, the aerospace industry and many others. And these 'actors' in turn are not genuinely unified, but are themselves fields of contention, where policy decisions emerge from previous bargaining sequences, between planetary scientists and the supporters of the manned programme within NASA, between the Air Force and the Navy within the Pentagon, between various aerospace corporations and so on. Similar complexities are present not only within a multinational body such as the European Space Agency, but even within such apparently homogenous entities as the Chinese and former Soviet space programmes.

Yet while there are similarities, there are also differences. History, culture, value systems and 'domestic' political structures also help to shape the objectives and modalities of space programmes. The Indian, Japanese and European programmes are all clear examples of the different ways in which such factors can structure a space effort.

Liberalism also lays stress on the way in which national policies adapt to the international environment during the complex bargaining sequences that are a feature of the international space regime, and can be seen in the governance of the geostationary Earth orbit, and in the development of the International Space Station. Liberal international relations theory has seemed particularly appropriate since the end of the Cold War. For the first three decades of the space age, space was dominated by the politico-military confrontation of the two superpowers. And, notwithstanding the stunning achievements of their manned and robotic space exploration mission, superpower use of space was itself dominated by military applications. Because of this, the end of the Cold War, which coincided with the emergence into maturity of the programmes of other actors such as Europe, China, India and Japan, led to optimistic speculation that a new era of space utilisation was dawning, whose tone and content would be overwhelmingly civilian.[31] Certainly, the gradual emergence of a growing multipolarity is evident, and it is noteworthy that the key actors involved, Europe, China and India, all have vigorous space programmes.[32]

During the Cold War these space powers tended to justify their programmes in terms of non-military criteria such as economic development, technological progress, communications advances and environmental surveillance. In the past decade, however, the military (as well as the broader security) rationale has been advanced without embarrassment. The utilisation of space has thereby taken on a characteristically mixed character in which it can still be argued that traditional power considerations remain prominent. Space has proven to be a domain where non-military facets of power can be exploited, and which are sometimes particularly advantageous.[33] But the end result can still be viewed as a struggle for power and influence in the global system using space as an instrument or utility. In the utilisation of space, 'we have witnessed a trend away from competition for prestige, with ulterior motives of a markedly military nature, to a scientific and economic competition, coupled with a military reality'.[34]

However, it should be remembered that this is not the first time that an apparent evolution from a realist to a liberal interpretation of space international activity has been celebrated. The superpower competition entered a period of improved relations in the first half of the 1970s that was characterised as *détente*. The willingness to explore potential areas of competition that *détente* represented was visible in the increased interactions of the two states' space programmes. The Apollo crews that landed on the Moon between 1969 and 1972 brought back considerable quantities of lunar soil for scientific analysis. In 1970 the Soviet Union performed the same feat with the unmanned Luna 16 spacecraft which also successfully returned lunar samples. Subsequently the two countries exchanged samples from the different lunar sites they had explored, evidence of the move from a purely competitive, to a cautiously cooperative superpower relationship. This move lends itself to interpretation through a liberal pluralist rather than a straightforward realist perspective.

An even more dramatic example of this new approach was the political symbolism constructed by the Apollo–Soyuz mission of 1975, in which the two countries docked their spacecraft together in orbit for the first time, with the two crews visiting each others' spacecraft. Both superpowers also cooperated with other countries in the years that followed. However, the superpower space cooperation was unable to survive the re-emergence of Cold War antagonisms at the end of the 1970s.

Liberal interpretations of international relations are also more useful in explaining those dimensions of space policy at the confluence of domestic and international politics, seeing no sharp boundary between the two. In the early 1960s, for example, the vigorous American space programme was being driven both by a domestic requirement of the Kennedy administration to divert attention from other set-backs such as the Bay of Pigs disaster, and a desire to use the exploration of outer space as a tangible manifestation of the meaning underlying the President's frequent rhetorical references to the existence of the 'new frontier' for the American people. At the same time, it was also a reaction to the successes of the Soviet space programme and reflected a perceived need to demonstrate American strength to an international audience of nervous allies and uncommitted Third World states.

The neoliberal approach has clear utility in terms of the analysis of international organisations and of space-related regimes. A major driver of the development of neoliberal theory in the 1980s was the demonstrable fact that 'levels of international cooperation were much higher than could be explained by neorealist theory'.[35] While agreeing that states were the key actors in international relations, neoliberals suggested that international organisations played a crucial role in the international system, particularly in the operation of international regimes, defined as 'sets of principles, norms, rules and decision-making procedures'. The neoliberal rapprochement with neorealism produced a view of international relations in which states sought to develop and enforce international regimes based on rules and law, even while 'political outcomes continue to be heavily influenced by power politics'.

The efforts by the West European states to coordinate their space programmes within ELDO and ESRO in the 1960s and ESA from the early 1970s provide interesting case-studies for the neoliberal approach, and these organisations would benefit from detailed investigation along these lines. The European Space Agency to date has operated through the *harmonisation* of national policies rather than through their *integration* within a supranational framework. As an organisation it therefore lends itself to an analysis reflecting neoliberal assumptions about the centrality of state policies in explanations of international cooperation and the strategic decision-making processes that drive them.[36] Certainly in the post-Cold War period international cooperation has been one of the dominant themes in space, with the construction of the International Space Station as its most dramatic symbol. Liberalism has some purchase in explaining particular aspects of space history, notably international cooperation in space, and the European experience in combining space exploration with the broader socio-political goal of European integration.

The utility of liberal analysis for explaining some aspects of space policy is not diminished by the emphasis on military issues in contemporary debates about space policy. Since the mid-1980s, the myth of 'space demilitarisation' has moved away from the idealist conception of a total ban on military space systems, to a more neoliberal account aiming at constraining the military use of space by promoting inter-state cooperation and commitment towards regulatory treaties.[37]

The space arms control regime is currently inadequate and is in need of urgent upgrading, especially under an emerging multipolar system.[38] The creation of such a regime poses exceptional difficulties however, given the variety of possible anti-satellite operations, and the overwhelming overlap between the civil and military technologies in this field.

The regulation of space reaffirmed that while international law exists in part to prevent the abuse or indiscriminate use of power in the world, it cannot be made effective without the possibility of power being exercised. For Klaus Knorr, there are two primary ways in which states came to organise their space activities in compliance with the emerging international law of space. One is through the enforcement of law by a supranational body, in this case the United Nations Organization, with all its limitations as an intergovernmental organisation, the second is through state commitment to *positivist* international law and action,[39] reflecting inter-governmentalism and a system of checks and balances.

The dramatic dominance of the early space age by the United States and the Soviet Union was a reflection of the broader historical shift in geopolitical power that had been inaugurated by the Second World War. That conflict saw the end of a long historical period of European domination of international relations. Neither of the two superpowers that pioneered the human entry into space were fully European. The Soviet Union straddled Europe and Asia, while the United States was an entirely non-European power. As with many other aspects of international relations, this reflected the way in which the political centre of gravity had moved away from the European powers, who had been devastated by the two world wars and relative economic decline. When the European powers did enter the field, they would do so collectively, again reflecting the broader currents of European integration that were to be such a feature of the subsequent half-century, and which were central to the development of liberal international relations theory in the 1960s and 1970s.

Liberalism is also relevant to debates over the relationship between globalisation and space policy. Liberals have emphasised 'the increasing irrelevance of national borders to the conduct and organisation of economic activity'.[40] The space age saw the full Earth globe become an iconic image for humans in a way that had never previously been the case. World-circling satellites and spacecraft annihilated distance and demonstrated the insubstantiality of international frontiers. The space age showed repeatedly that many of the political, economic and social issues of the Cold War period could neither be contained by, nor successfully resolved within the constraints imposed by the traditional boundaries of nation states. At the same time, the growing cooperation between different national space programmes emphasised the inadequacy of traditionally autonomous states in addressing certain contemporary challenges. The high cost of space ventures, along with the associated technological interdependence makes cooperation both necessary and inevitable.[41]

The fact that Outer Space has been a realm about which humans speculated and into which they projected their beliefs long before humanity's physical movement beyond the confines of Earth is important in terms of *social constructivist* approaches to understanding international relations. For social constructivists like Onuf, international relations is 'a world of our making'.[42] There is an external objective reality, composed of mountains, seas, deserts, rainfall and so on, but the social world that people inhabit, of tribes and states, economic and political institutions, ideas, norms and cultural values and so on, is a social construction. It is created by dialogue and intersubjective consensus that produces provisional agreement on both what constitutes the external reality, the ontological environment in which humans find themselves, and also on the meaning of all or parts of that 'reality'. Such a consensus is subject to change, both evolutionary and, occasionally, revolutionary. Physical reality may be prior to human intervention, but it is human beings who give *meaning* to the reality they encounter and relate to it in terms of that meaning. When such interpretations stabilise, it is as the result of social processes, and the same processes may challenge that consensus in the future. Ideas and understandings shape reality and have the capacity to change it, as evidenced by the contribution made by Gorbachev's 'new thinking' on international relations to the ending of the Cold War.[43]

Agreed meanings emerge from a contested intellectual environment in which the interests of the antagonists are central in their attempt to define reality. The acquisition of knowledge itself 'is a societal process, based on incentives, motives and interests of individuals in a natural and societal environment'.[44] Thinking in these terms is important when trying to understand space politics. Decisions for and against space-related policies, and even decisions about whether to have such policies, are shaped by world views and beliefs about what space does, or might, represent. This can be seen in debates over whether to allow weapons to be placed in space, or what sort of regime should govern human activities on the Moon.

In this regard, post-structuralism would seem to have a particularly useful part to play in the analysis of the international politics of space. This is not simply due to its function as a critique of alternative conceptions such as realism. The critique of modernity as such has a particular resonance when dealing with the technocratic ambitions of space programmes. Modernity seeks to reveal the mysteries of the universe through the application of human reason. It sees history in terms of a linear progress towards a distant but real *telos* of greater understanding and material well-being. It seeks to shape the future 'through powers of scientific prediction, through social engineering and rational planning, and the institutionalisation of rational systems of social regulation and control'. [45] The space programmes were, and are, an apotheosis of this mode of thinking.

The critique of modernity by post-structuralism is therefore particularly appropriate in considering the various claims made on behalf of space exploration and utilisation, of deconstructing the processes by which certain ways of thinking about space emerged and became seen as valid, while others did not. It is in the world of ideas that post-structuralism provides the greatest purchase. Post-structuralism contests the idea of rationally derived, incontestable social or scientific truth. From a post-structuralist perspective, action takes place within a pre-existing structural and narrative framework. This structure in turn sets limits as to what is considered possible.

At the functional level the postmodern world is an age of compressed space and time. Satellite technology and the looming menace of nuclear-tipped long-range ballistic missiles have helped to produce a world where flows of information, capital and ideas are almost instantaneous, while trade, military power and populations move about the world at undreamt of speeds. Again, this is an area where post-structuralist approaches to the study of international relations are particularly relevant, as is the inside/outside distinction between domestic and international politics, which plays out somewhat differently in the inside/outside issues of post-sovereignty represented by human activities beyond the Earth's atmosphere.

Gender theory is also an approach from which the study of space policy would benefit, both in terms of including gender analysis as an approach and of making women's experiences part of the subject matter. Women were strikingly absent from the early superpower space programmes, the flight of Soviet cosmonaut Valentina Tereskova in 1963 being the exception that proved the rule, and only in recent years has the subject of women in space begun to be systematically analysed. The gendered nature of early astronaut selection is all the more striking given the public image of space exploration as representing the best of humanity, at the cutting edge of progress. It took 20 years before the first female American astronaut followed her male compatriots into space, and a further 20 years elapsed before a female astronaut was commander on a space mission.[46]

In the dramatic era of the superpower space race, there were plenty of women who met most of the NASA selection criteria, and the US Air Force even initiated a programme to identify suitable female astronaut candidates. However, in December 1959 the programme was cancelled. Potential female astronauts were 'the right stuff, the wrong sex'.[47]

Just as there are numerous international relations theories that can provide purchase on the international politics of space, there are also a number of different ways of conceptualising space that have been significant since the 1950s. Space has typically been viewed in terms of three perspectives; as a sanctuary, as an environment and as a theatre of war. This is a useful way of thinking about space as a whole and in addition, the three categories have some intellectual correlation between IR approaches characteristic of post-structuralism, liberalism and realism.

The *exploration* of space can be seen as being of value to all humanity, not only in terms of the technological advances required in order to be able to explore it, and the 'spin off' from such technologies to other aspects of desired progress, but also because unexplored space represents a dominion of unclaimed knowledge in itself.[48] Since ancient times humans had wondered about the nature of space and looked to it as a source of explanation and prediction of terrestrial unknowns.[49] Space in this sense is a value, 'the final frontier', a realm to be explored for the secrets it can slowly reveal to humanity. In 1958, The Majority Leader of the US Senate, Lyndon Johnson, called for the United States to use international space cooperation to build confidence and peace between the nations of Earth, declaring that it would be appropriate for the US to propose to the member states of the United Nations that they join the US in 'this adventure into outer space together. The dimensions of space dwarf our national differences on Earth'.[50]

The *exploitation* of space refers to the actual use of space as another political theatre, where states in the long term might seek to exploit cosmic resources for their power potential, but in the short term exploit space because of its ability to produce 'force multiplier' effects on their existing terrestrial military capabilities,[51] or as an economic asset.[52] Here, space is simply a medium for the acquisition or exercise of power, strategic, economic, ideological, but always with profound political implications. This dichotomy was present from the beginning of the space age. It can be seen in the schizophrenic attitude of the German and Soviet rocketry pioneers in the interwar rocket societies VfR and GIRD who had scientific and spaceflight-ideological goals favouring space exploration, but had to subordinate them to the demands of the military.

This dichotomy between competition and cooperation is of great political significance for the major powers in relation to their foreign policy interests in space activity. On the one hand they will cooperate where refusal to do so is likely to 'stimulate moves towards space independence by other nations'.[53] On the other hand, they wish to preserve their hegemonic position unchallenged and will therefore restrict cooperation to those areas of space activity that will not affect the stable and predominant position enjoyed by the major space powers, particularly when they fear the transfer of technology with critical military or commercial value.[54]

The continuing centrality of this dilemma can be seen in United States' space policy, which has become focussed on the requirement to gain and maintain 'space control'. The United States has edged steadily closer to the acquisition of the military and infrastructural capabilities necessary to conduct space warfare and ensure that the US is able to exercise an effective monopoly of military space use in wartime. But the United States has so far resisted the temptation to cross the threshold of space weaponisation and move swiftly to deploy such capabilities because, so long as America's hegemonic position in space is under no significant challenge, then it is preferable to maintain the current cooperative and non-weaponised space environment, since it meets all the United States' requirements.[55]

From the dawn of the space age space has posed a challenge for those seeking to create policy for it. Because it was a completely novel theatre of political activity, the temptation was to use terrestrial analogies to understand it. This was understandable, and has significantly shaped subsequent conceptualisations of outer space as a realm of political activity, but it has also often been misleading and unduly constraining in terms of conceiving of what might be possible on what has been called the 'final frontier'.

Chapter 2

Propaganda and national interest

Scientific socialism and the Soviet space programme 1957–69

Power is an essential ingredient for effective action in international politics. As a phenomenon, it combines all the components of national strength, tangible and intangible, real and potential, into a unity. The measure of power determines the extent to which a state can exert influence in global politics; it determines the capacity to influence, to manipulate and to control.

Power is often understood as being synonymous with the capacity to threaten or exert force, particularly military force. However, power defined as the ability to influence outcomes in a desired direction is a more complicated and nuanced phenomenon than simply the ability to impose one's will by force. With states, as with individuals, the 'charismatic appeal of authority' is also important, and a reputation for power or perception as being strong, can also be significant. It was recognised early on that in the peculiar circumstances of the Cold War, the superpower struggle for power would be very much a struggle for the control of people's *minds*, of their perceptions of reality.[1] In this battle over perceptions, propaganda was central. Propaganda can be defined as the use of symbols in an effort to manipulate the beliefs, attitudes or actions of other people, or to propagate a particular doctrine or practice. Seen in this light, it is easy to appreciate the symbolic importance of activities in space. Being at the cutting edge of technology, achievements in space can be presented and interpreted as a symbol of human progress, and a validation of a particular social and economic system.

The realist scholar Morgenthau, for example, pointed to the importance of *prestige*, which he defined as a 'reputation for power', and which could be a tool for achieving larger political goals. When states pursue policies designed to increase their prestige, they are seeking to confirm an evaluation of strength, excellence, even superiority. In Karl Deutsch's terms, 'prestige is to power as credit is to cash'.[2] Recognition of these elements may be sought domestically, particularly when national morale is deemed to be low after an historical reverse, but it is their recognition and acceptance externally by other actors in the international system that is most important.

The acquisition of prestige sustains and reinforces a reputation for strength and can contribute significantly to a state's authority in global politics. The early superpower space race is a dramatic example of this in practice, with both states seeing their international authority as being critically affected by domestic and particularly external perceptions of their relative performance in space. Soviet space policy from the outset sought to exploit the programme for military and scientific benefits, but also, crucially, for political gain.[3] The space programme was so dramatic and so compelling that it fulfilled the purposes of Soviet propaganda with unusual effectiveness. K J Holsti noted that Soviet interests could be promoted through 'propaganda programmes that would bypass foreign governments and influence foreign populations instead. These populations, it was hoped, would in turn force their governments to act in a manner consistent with Soviet interests'.[4] The space programme could do this effectively because, as George Allen, Director of the US Information Agency noted in the aftermath of the Sputnik launch, space had 'become for many people the primary symbol of world leadership in all areas of science and technology'.[5] The space programme would demonstrate the clear existence of a modern scientific, technical and industrial base in the Soviet Union.

This was important both in the positive sense of promoting a particular image of Soviet communism and the USSR, and in the negative sense that it was needed in order to counter the message of Western propaganda, which portrayed the Soviet Union as an introspective and backward state. A successful space programme would validate Soviet claims about the effectiveness of their system and its superiority over western capitalism. The space programme therefore became inextricably intertwined with the Soviet propaganda machine, even recruiting former party propaganda specialists such as Zimyanin and Tyazhelnikov into its ranks.

From a realist perspective, it was clearly the pursuit of power that encouraged the Soviet Union (and the United States) to seek to achieve advances in space exploration in the hope of increasing their military/ technological capabilities and prestige. Both sides increasingly recognised the dangers inherent in the nuclear confrontation and the fact that a full-scale nuclear war would represent mutual annihilation. Increasingly therefore, while continuing to seek advantage and superiority over their rival, they looked for other options to demonstrate their claims to superiority and the space race became an important surrogate for war. A crucial development in this regard was the doctrinal shift that had taken place at the Twentieth Congress of the Communist Party of the Soviet Union, where the CPSU had embraced the idea that a full-scale war with the West was not inevitable.[6]

But the space race was also a battle of images and perceptions, confirming the centrality of the ideational dimension of international politics. It is important in this context not to underestimate the symbolic and idealistic impact of the early space missions, particularly the manned flights. These were achievements that humans had dreamed about throughout recorded history, but had never previously come remotely close to attaining. Ericke, one of the German V-2 scientists who had followed von Braun to the United States, could describe the flight of Yuri Gagarin in 1961 as representing 'the height of human thought ... it not only imparts dignity to the technical and scientific aspirations of man, but also touches the philosophy of his very existence'.[7] Achievements in space in this period were therefore effective not only in demonstrating superiority in relation to the other superpower, but also in terms of identifying each state's historical project with the larger dreams of humanity.

For the Soviet Union, this was particularly true because of a number of ideological features that helped to shape Soviet self-perception and foreign policy in the middle of the twentieth century. These included the idea of 'scientific socialism', the 'correlation of forces' concept and the central position of propaganda as a Soviet foreign policy instrument.

The ideology that guided the policies of the USSR was that of Marxism-Leninism as developed by their Soviet successors. Marx had called his social theory 'scientific socialism' in order to distinguish it clearly from what he considered to be the rather utopian and insufficiently rigorous ideas of other socialists such as Saint-Simon and Fourier. For Marx these alternative socialisms described a superior future, but failed to critique methodically the existing social order and expose its failings in a scientific and methodical way. Marx in contrast, believed that he had successfully used the scientific method in vogue in nineteenth century Europe to discover the laws of history.

For this reason, and because it needed to industrialise rapidly and demonstrate quickly the superiority of socialism over capitalism, the Soviet Union placed enormous emphasis on the importance of science as a vehicle for progress and as a symbol of the superiority of socialism. The successes of the Soviet space programme were literally a gift from heaven therefore, enabling the Soviet Union not only to demonstrate to its own people and to the world the path-breaking achievements of Soviet science, but to demonstrate also the superiority of the ideological system that had made these achievements possible, scientific socialism. Though Marx had meant by 'scientific' socialism the nature of the methodology he was using, for the Soviet leadership, the relationship worked both ways, with the successes of Soviet science validating socialism as practice, just as the use of the scientific method had validated it as theory. This logic is clearly reflected in the claim by Soviet Premier Nikita Khruschev that the launch of Sputnik, the world's first artificial satellite in October 1957, had been carried out with the aim of 'convincing the people of Russia, China, India as well as Europe that our (communist) system is the best'.[8]

A second important ideational factor was the correlation of forces concept. Western security analysts viewed the Cold War struggle through the prism of the balance of power concept. Though the nature of power was rarely defined by balance of power theorists, it is clear from their writings that what was meant was military power.[9] Military power is clearly important in understanding the overall pattern of influence in international relations, but it is not the only form of power. In the Soviet Union, the balance of power concept was dismissed as being based on a false view of the international system. The preferred Soviet concept of the 'correlation of forces' was seen as being a broader and more subtle concept than balance of power. In estimating the relative balance of influence between east and west it included the military dimension, but also took into account the overall class, social, economic, political, ideological, ethical and other forces operating within and between the two opposed alliance systems.[10] Military power was seen as having become relatively less important compared to economic and socio-political factors. In addition, the struggle was no longer simply between rival states, but between groups of states, international movements, classes, popular masses and parties.

The Soviets criticised the 'two bloc' variant of balance of power thinking both because it did not take into account the social nature of the rival forces, and because it did not reflect the Soviet view of the protagonists as representing the forces of progress versus the forces of reaction in a competition whose ultimate outcome was historically determined by the processes Marx had revealed through scientific socialism. Soviet writers also criticised the balance of power concept because of its reliance upon war as an instrument of policy. In the era of nuclear weapons this was seen as being a desperately dangerous strategy. It was argued that the very fact that nuclear weapons were too dangerous to use in the superpower struggle served to raise the significance of the non-military elements in the correlation of forces, such as economic, political and ideological factors.

It is not difficult to see why the startling success of the Soviet space programme after 1957 lent itself easily to incorporation into both the promotion of scientific socialism and the correlation of forces equilibrium. Because the correlation of forces incorporated all aspects of the international competition, the symbolism of the Soviet space triumphs clearly contributed to the overall calculus of power from the Soviet perspective.

The Soviet space programme between 1957 and 1991 can be divided into two fairly clearly defined phases. Both were characterised by the energetic use of space achievements to underpin the effectiveness of propaganda as a Soviet foreign policy instrument. But the central propaganda theme of the two phases was significantly different. In the first phase, from 1957 to 1970, the emphasis was on competition with the capitalist world in general and the United States in particular. In the second phase, from 1971 to 1991, the space programme was used as an exemplar of the virtues of *détente* and international cooperation, and evidence of the desire of the Soviet Union to work together with all 'peaceloving' states. But the underlying message in both periods was the same, the inherent superiority of the Soviet system over its capitalist rivals.

The years from 1945 to 1957 were an era of acute confrontation between the USA and the USSR. The global clash of interests between the superpowers created situations pregnant with the possibility of thermonuclear war and a general climate of tension and unrelenting pressure in international relations. By the late 1950s, however, two general tendencies were beginning to moderate this situation. One was the sobering realisation that a nuclear war might well mean total and mutual nuclear annihilation. The second was a growing diversity in the international system as a result of decolonisation and post-war recovery, which gradually brought an end to rigid bipolarity and an increase in political pluralism.

These two changes meant that for the contending superpowers, there was now an audience to be won over, composed of wavering allies and the non-aligned, and also that new symbols and pressures would be required that did not involve the risk of nuclear war. The surrogate superpower competition represented by the space race answered both these requirements.

The Soviet Union was well placed to exploit this opportunity at the end of the 1950s. The USSR had begun developing long-range ballistic missiles immediately after the end of the Second World War. Although the Soviet Union exploded its first nuclear weapon in 1949, it had no long-range bombers capable of delivering these weapons to targets in the United States. The solution was the development of long-range missiles. Stalin was a firm believer that long-range missiles would be the decisive weapon of the future and initiated the missile programme that was brought to fruition under his successor, Nikita Khruschev.[11]

The Soviet Union began this programme with some advantages. There was a tradition of interest in space exploration in the Soviet Union dating back to the work of the nineteenth century Russian pioneer, Konstantin Tsiolkovsky.[12] Tsiolkovsky's work, particularly his early 1930s studies on the idea of a multi-stage rocket, was an important platform for the ballistic missile programme that would place a satellite in orbit two decades later.[13] In addition, the Soviet military had favoured the development of battlefield rocket systems in the build-up to the Second World War. They were thus open to the possibilities for using missiles for military purposes. In the 1930s an unofficial group of scientists in the GIRD group had studied the problems associated with developing long-range rockets and had carried out some small-scale launches. Their status changed significantly when the Soviet Armaments Minister Mikhail Tukhachevsky gave them official endorsement and government funding for further research and development. Crucially this development tied the subsequent Soviet rocket programme into the military establishment, where it remained until the end of the Cold War.[14]

All the allied powers had noted the effectiveness and potential of the V-2 missiles used by Germany in the closing stages of the war. The Allies had agreed to share the German V-2 materials that would be captured at the end of the war, but in practice Britain, the USA and the USSR simply seized what they could and shared little or nothing. The drift towards Cold War had already begun and the military potential of the V-2 technology had been clearly demonstrated in the last two years of the war. As the Soviet forces closed in on Germany in the final months of the war, they overran the main German rocket development facility at Peenemunde on the Baltic coast. However, von Braun and the other leading German engineers had already fled, and the facilities themselves had been almost totally destroyed by RAF bombing and the retreating Nazis. Nevertheless, the Soviets did capture many of the lower ranking engineers, as well as crucial lists of V-2 components and their suppliers. The captured engineers and surviving V-2 hardware were shipped back to the Soviet Union. The knowledge and technology so gained proved crucial when added to the pre-existing Soviet research and development effort in military rocketry.[15]

The driving force behind the Soviet development of long-range rockets after 1945 was the military requirements produced by the Cold War confrontation with the United States and its allies. From the outset, the pursuit of scientific knowledge for its own sake played little if any part in the motivations for the programme. Propaganda would become central from late-1957 onwards, but the one constant was the military rationale.[16] The military requirements of the USSR underpinned the space programme as indeed they did all aspects of the rapid Soviet advances in science and technology after 1945.[17] An example of this synergy was the fact that the government decision to initiate research into satellite development came three months after the decree authorising development of the R7 ICBM launcher, which would also subsequently launch the satellite.[18]

The emphasis in the Soviet Union at this time was on the production of long-range missiles for military purposes. The USSR had developed the hydrogen bomb in 1953, but the weapon weighed two tons and to deliver this against targets in the United States would require a rocket three times as powerful as the largest American rocket under development. The Soviets, under the leadership of 'Chief Designer' Sergei Korolev, developed a rocket codenamed R-7. It was small and clumsy and required 20 engines to fire simultaneously, but test launches in early 1957 demonstrated that it worked.

It was at this point that a number of political dynamics converged to launch humanity into space. Soviet leader Nikita Khruschev was under intense political pressure at home and abroad in early 1957. His position was still politically fragile despite having recently emerged victorious from the power struggle that had followed the death of Stalin in 1953. The outbreak of the Hungarian rising and its subsequent violent Soviet suppression in 1956 had badly damaged the Soviet Union's international prestige. Poor harvests at home likewise weakened Khruschev's political position. A positive propaganda coup would clearly be welcomed in such circumstances. In the summer of 1955 Korolev had pointed out that a satellite launch 'would have enormous political significance as evidence of the high development level of our country's technology'.[19] In addition, Oberg suggests three other factors that influenced Khruschev's decision to support a satellite launch. In a general sense, a successful launch would signal to disaffected political elements within the USSR that Khruschev really was leading the country to a glorious future. Domestic political considerations appear to have been at the heart of the decision to approve the Sputnik launch, rather than any well-conceived idea to challenge or upstage the United States. Second, it would overawe the traditionalists in the Red Army who were opposed to Khruschev's plans to reorganise the armed forces radically, and in fact in the aftermath of the successful Sputnik launches Khruschev 'virtually rammed rockets down the throats of Red Army traditionalists'.[20] Finally, it would demonstrate the existence of a long-range missile system and thereby contribute to the Soviet ability to deter an American nuclear attack.[21] This was a crucial consideration given the perennial Russian sense of insecurity, and entirely reasonable given her experience of having been attacked from the West twice in the previous 50 years.

The centrality of the military rationale, and the overlapping nature of the missile technology in the military and space fields were crucial. It is questionable whether the USSR could have afforded to fund both areas on such a lavish scale had there not been major synergies between them. In the event of competition for funding, the defence sector of necessity would always have been given preference.

Korolev had been arguing since 1954 that it would be possible for the Soviet Union to orbit a simple satellite, launched on a modified R-7.[22] Eager for some success that would demonstrate Soviet technological capability, Khruschev approved the proposal, with the proviso that the launch should take place in October 1957, the fiftieth anniversary of the Russian Revolution.

The short deadline meant that plans to launch a satellite carrying sophisticated instrumentation were abandoned in favour of a simple battery-powered short-wave radio transmitter, which was launched on Sputnik, into an orbit that carried it over Europe and the United States, where its distinctive signals could clearly be picked up. Sputnik had a powerful and largely unanticipated effect. Not since the Japanese attack on Pearl Harbor had Americans felt so vulnerable to a foreign power. The Sputnik launch triggered an outburst of American self-criticism and even self-doubt, and its effect was increased by the sensationalist reporting of the American press and the many confused and contradictory attempts by the Eisenhower administration to minimise its significance.[23] While some Americans attempted to play down the Soviet achievement, most felt that the USSR had achieved a tremendous propaganda coup, and that America had been humbled.

It was assumed that if the Soviet rocket could lift a satellite into orbit, it could just as easily carry a nuclear warhead from the USSR to the USA. Sputnik was so much larger than the American Vanguard satellite under development, that it was assumed that Soviet missiles would be able to carry much larger nuclear warheads than their smaller American equivalents.[24] The launch encouraged the United States to overestimate greatly the military capabilities of the Soviet Union and the spectre of the 'missile gap' would haunt US administrations until well into the 1960s – ironically until evidence obtained from American satellite reconnaissance gave them more accurate information. In reality the R-7 rocket was not an ideal ICBM launcher, and was only being produced in very small numbers. But US intelligence was unaware of this fact and assumed that production levels were very high. It was significant however that the first successful R-7 ICBM launch, in August 1957, was described in surprising detail in a communiqué by the Soviet Press Agency, TASS. TASS strongly emphasised that the Soviet Union now had an effective ICBM. The American press and government paid no attention to the announcement. It took the drama of Sputnik to seize their attention.[25]

After the astonishing news of the Sputnik launch, President Eisenhower attempted to calm American anxieties by arguing that the US satellite programme had 'never been conducted as a race with other nations'.[26] This was essentially true, but completely missed the political point. Eisenhower insisted that 'one small ball in the air does not raise my apprehensions, not one iota. We must ... find ways of affording perspective to our people and so relieve the current wave of near hysteria'.[27] Eisenhower was in fact right to argue that the American people were overreacting, but the hitherto prevailing perception that the Soviet Union was a clearly backward society in comparison to the United States made its space achievement seem all the more startling and alarming. In 1956 and 1957, US analysts had reported over a dozen announcements in the Soviet press on the plan to launch a satellite during the International Geophysical Year, a period of international scientific activity covering this period. These reports gave considerable detail on the satellite and its intended orbit. The fact that the launch took the American public completely by surprise was not the result of efforts by the Soviet Union to maintain the secrecy of the project, but rather 'to the arrogance of American mass media and to Cold War paranoia'.[28] The previously implausible claims of Soviet propaganda now seemed disturbingly validated, making subsequent Soviet claims more difficult to dismiss.[29] For the USSR, the great achievement of the Sputnik launch and of subsequent space spectaculars, was that it raised the possibility that 'success in space implied superiority on earth'.[30]

At first Soviet Premier Khruschev himself had not realised the full implications of the achievement. He congratulated the scientific team over the telephone and then went to bed. It was only when he was shown evidence of the reaction in the West that he understood the nature of the propaganda gift he had been presented with.[31] While Khruschev had not anticipated the scale of the propaganda benefits that the Sputnik launch would bring to the Soviet Union, he quickly saw how Sputnik and its successors could be used to demonstrate to the world that the USSR was a highly advanced society, capable of competing with the West on equal terms. Certainly this was how it was perceived in much of the West. The *Manchester Guardian* editorialised that 'it demands a psychological adjustment on our part towards Soviet society'.[32] Opinion polls by the western media taken in the immediate aftermath of Sputnik indicated that clear majorities in Italy, France and Britain were now of the opinion that the USSR was ahead of the United States in terms of scientific development.[33] For Khruschev therefore, there most certainly now was a 'space race', just as there had hitherto been an arms race. And from the beginning, the two would be closely linked. Khruschev had intuitively understood that the American people were vulnerable to a 'Pearl Harbor syndrome', a deep-rooted fear of being taken by surprise by an enemy whose actions they could not adequately monitor or predict. He was quick to play on these newly aroused fears by misleadingly claiming that the Soviet Union was manufacturing ICBMs like sausages on a production line. It was a linkage that he was to repeat both for the domestic and international audiences in subsequent years. In a speech in 1959, for example, he declared that 'the Soviet Union has rockets in a quantity and of a quality unequalled by any other country. This can be confirmed by the launch of our Sputniks and cosmic rockets'.[34]

In reality, however, Soviet long-range missiles were limited both in quantity and effectiveness, but the Soviet Union was understandably intent on convincing the USA that the opposite was the case. This crucial deterrence bluff was made possible by the impression created by the space launches of a dynamic, well-funded and technologically sophisticated rocket programme in the USSR. Having created this impression, however, the USSR now had to maintain the image of dynamism and innovation that international opinion had ascribed to them.

The impact of the Sputnik launch was rapidly consolidated with the launch of Sputnik II. The second launch reinforced American concerns, since the new satellite was five times the size of its predecessor, carried a living creature on board (a dog called Laika) and was launched only 30 days after the first satellite had been orbited. Laika died of heat exhaustion after a week in orbit. Nevertheless, the flight had demonstrated that living creatures could survive beyond Earth, and that it would be possible for human beings to make spaceflights. This was a crucial development, because as subsequent decades would show, 'to the layman the true conquest of space was represented above all by manned spaceflight'.[35]

The second launch confirmed the Soviet belief in the power of the space programme as a highly effective propaganda weapon in the Cold War. Khruschev's political support became crucial to the resources that the programme received over the next few years. But this support came at considerable cost. For Korelev, the head of the Soviet space programme, the purpose of space launches was the scientific exploration of the solar system, leading in time to the permanent presence of human beings in space and on other bodies. For Khruschev however, it was simply a propaganda tool, with no significant value in itself. In the years 1958 to 1961, the United States launched 56 satellites, while the Soviet Union only launched 12. Unlike the American launches, few of the Soviet missions produced significant gains in scientific knowledge.[36] For Khruschev, space research for its own sake was of no interest, and follow-up missions were invariably cancelled because they might appear repetitious, even though they would have been scientifically important in confirming and developing knowledge. Instead, Korolev was forced to conform his space launch schedule to Khruschev's diplomatic agenda.

This would have damaging medium-term repercussions on the direction and nature of the Soviet programme. While Korelev wanted to see a steady and coherent pattern of launches that would consolidate knowledge and gradually develop increasingly sophisticated technology capable of achieving more ambitious goals, Khruschev simply wanted a series of 'firsts' that would reinforce the sense that the United States was lagging behind the Soviet Union in terms of advanced technology. An example of this was the development of re-usable spacecraft. Work in this field began in the United States in 1962 and would ultimately produce the Space Shuttle. In the USSR, pioneering scientists such as Tsiolkovsky and Zander had speculated on such vehicles as far back as the 1920s, but Korolev's proposal to initiate a programme for their development in the 1960s was dismissed.[37] Under pressure from Khruschev therefore, Korolev was forced constantly to adapt existing technology rather than being allowed to develop the next generation of technology – an ultimately self-defeating strategy for the USSR as regards the Moon landings.

In a real sense the use of the phrase 'Soviet space programme' to describe the USSR's efforts at this time is misleading. There was no such programme in the sense of a single integrated organisation working to achieve a coherent set of objectives that were part of an overall plan. Soviet launches were achieved by design bureaux, often fiercely competitive with each other and the Soviet leaders reacted to their proposals in a fairly ad hoc manner, influencing the choice of launch dates and exploiting propaganda benefits, but otherwise having no clear agenda or direction as far as the development of space technology was concerned.[38]

In the short-term it was an effective strategy, however. In 1958 the USSR began launching spacecraft towards the Moon. Luna 3, launched in October 1959, was able to produce photographs of the far side of the Moon, the side that is never seen from Earth. The USSR, as discoverer, was in a position to name the new features discovered, and rubbed in its propaganda success by giving them designations such as 'Mount Lenin' and the 'Sea of Moscow'. In August 1959 the Soviet Union achieved another first by launching dogs into orbit on Sputnik 5, and returning them safely to Earth.[39]

That this succession of achievements was having the desired effect is illustrated by the comment of the new NASA Director, that, 'they hope to become so superior in their scientific capabilities that they will win world domination through industrial power rather than through shooting wars'.[40] Even in regard to the latter consideration the Soviet achievements were alarming. NATO believed itself to be heavily outnumbered by the Soviet and other Warsaw Pact countries in regard to conventional armed forces. In order to offset this perceived disadvantage, it had developed a strategy that depended upon the technological superiority of NATO military systems, and the early use of nuclear weapons. The Soviet space achievements therefore put in question the assumptions on which the whole of NATO strategy was based, making its deterrence policy seem far less convincing. For some they had in fact brought about a revolutionary change in the international system, acting as 'the path by which the Soviet Union has taken a shortcut to global power status'.[41]

In the first half of 1960 the Soviet leadership began to take a much greater interest in the political and military possibilities offered by the space programme.[42] There is evidence to suggest that this was largely due to a perceived need to compete effectively with the accelerating American programme rather than any new recognition of the value of space exploration.

Khruschev was quick to tie the success of the space programme to the power of communist doctrine. Gagarin's successful manned flight in 1961 was trumpeted as demonstrating the correctness of Marxist–Leninist ideology.[43] Gagarin himself publicly credited the Communist Party for the successful flight.[44] The even more optimistic cosmonaut Popovich declared before his launch in 1962 that he was 'blazing a trail for all mankind to the communist future'.[45] Statements such as these were directed as much towards the Soviet Union's allies as they were towards the West. The space programme both reassured the Warsaw Pact allies that the Soviet Union could deliver on its promises of military and economic security, and warned the dissident communist former ally of China that the USSR was a superpower to be taken seriously, no 'paper tiger'.

Between 1961 and 1965, therefore, the USSR used the Vostok and Voshkod manned spacecraft programmes to achieve a series of 'firsts' that were often little more than stunts, though several of them also had a purpose in demonstrating capabilities that would be required in meeting President Kennedy's 1961 challenge to become the first country to land human beings on the Moon. Vostok 1 saw the most dramatic achievement, in April 1961, with the flight of Yuri Gagarin, the first human to fly beyond the Earth. The humanistic symbolism of Gagarin's flight was all the more marked because it was followed immediately by the US Bay of Pigs fiasco, when American-backed insurgents were totally defeated in an attempt to overthrow Fidel Castro's government in Cuba. Gagarin's subsequent victory parade in Red Square was the first time that the communist authorities had allowed such an event to be broadcast live by the western media. Gagarin himself was awarded the Order of Lenin and then sent on a world-wide tour, acting as a publicist for the achievements of the Soviet Union.

Due to the international impact of Gagarin's flight and that of the second man in space, German Titov, the propaganda focus changed from unmanned satellites to manned spacecraft. Vostok 3 and 4 became the first spacecraft to 'rendezvous' in space, approaching within seven kilometres of each other in 1962. Vostok 6 saw the first woman to fly in space, Valentina Tereshkova. Tereshkova's flight allowed the USSR to suggest not only that it was technologically superior to the USA but that it was socially superior also. Her June 1963 mission lasted three days, by which time she had spent more hours in orbit than all the American astronauts put together. She was described as a typical member of the liberated Soviet working class, someone of 'impeccable proletarian heritage'.[46] More importantly, the Soviet Union argued that the mission clearly demonstrated that the USSR was a society marked by equality between the sexes, and that no role was denied to women on the basis of their gender. Although he had been opposed to the idea of a female cosmonaut, after her flight Korolev underlined the propaganda message by declaring that her flight was 'one of the most striking demonstrations of the equality of Soviet women'.[47] Former US Congresswoman Clare Booth Luce accepted the Soviet claim that it was evidence of the emancipation of Soviet women and declared that 'it symbolises to Russian women that they actively share (and not passively bask like American women) in the glory of conquering space'.[48] In reality there were no other female cosmonauts in training and in fact no Soviet woman would again fly in space for nearly 20 years. Nevertheless, the USSR hosted an international conference for women in the week after her return, in order to maximise the positive associations of the mission.

Voshkod I in 1964 carried the first three-man crew, while Voshkod II featured the first extra-vehicular activity, with Alexei Leonov's 'spacewalk' in 1965. This dramatic series of 'firsts' epitomises the early years of the Soviet space programme. The logic was that by carrying out these activities before the United States did, the Soviet Union would demonstrate to America and the rest of the world its continuing technological, political and economic superiority. Soviet-manned flights were used not only to project a positive image of the USSR, but to reduce the damage caused by setbacks in other areas. The launch of the second manned flight in August 1961, for example, was timed to offset to some degree the likely negative publicity that Khruschev anticipated from the construction of the Berlin Wall.[49]

Throughout the development of the space programme, Khruschev, while ensuring that he was always publicly associated with its successes, decreed that the identities of other key figures in the programme should not be revealed. Korolev, for example, was referred to only by title, 'the Chief Designer', never by name. It was crucial for Khruschev's political strategy that in the domestic political context, he alone should receive the credit for the successes of the space programme. It was equally important that set-backs and failures were hidden both from the Soviet population and the wider world.[50] Launches were announced only after they had been successful, the differing purposes of various satellites were cloaked under the common 'cosmos' designation, even the exact position of the launch sites was kept secret. Korolev in turn, struggled to obtain the support and resources needed to sustain a coherent manned space programme. In 1963 he sought to have the manned space programme placed under the control of the air force in an attempt to give it coherence and assured access to resources.[51]

In the event, the short-sighted constraints on the Soviet programme's ability to emphasise the development of a successor generation of technology meant that it was the United States, not the Soviet Union, that achieved the goal of the first manned lunar landing, in July 1969. In an attempt to minimise the propaganda damage of this failure, the USSR subsequently attempted to claim that it had never in fact been pursuing the goal of a manned lunar landing. Their actions at the time clearly contradicted this claim, however, and documents declassified after the end of the Cold War and testimony from participants in the Soviet programme of the period demonstrate that the 'Moon Race' was in fact very real.[52]

The Soviet Union in the late 1950s was well aware of its relatively inferior position in relation to the overall capabilities of the United States and its allies. In a position of relative weakness, the ability to take the lead in a technologically impressive area such as space research was crucial. It was barely a decade since the USSR had been devastated in the Second World War, suffering losses that dwarfed those experienced by the USA, which had suffered no similar destruction of its industrial capability. Unlike, for example, the development of the atomic bomb, in which the USSR had to respond to the USA by subsequently developing its own, space exploration had allowed the USSR to take the initiative and place the United States in an apparently inferior position. A successful space programme could be correlated with superiority in other sectors, such as industrial capacity and the status of women. It was not only the Soviet Union itself that stressed such linkages, for the Western media played a powerful role in relating Soviet space capabilities to other sectors of Soviet achievement, the military in particular, and this despite the degree of scepticism about Soviet accomplishments that was always present to some extent.

However, the failure to develop space technology in a systematic and sequential manner meant that the series of 'firsts' could not be indefinitely maintained. The ultimate failure of the USSR to maintain Khruschev's bluff was to have important negative consequences for the Soviet Union. A problem with the prestige strategy was that it was a 'highly perishable asset'. Prestige returns diminished as both the domestic and international audience came to take spectacular achievements for granted, to expect more and better continuously, and to recognise increasingly that any leads would only be temporary. As the dramatic Soviet achievements of the first few years were followed by steady and accelerating American advances, most states continued to accept the United States as militarily, technologically and economically superior to the Soviet Union in most areas of activity, and this despite the enormous resources invested by the Soviet Union after Sputnik in an attempt to live up to its own superpower image.[53]

The Soviet effort to maintain momentum was no easy task, given that the Soviet economy was barely beginning to recover from the enormous losses inflicted by the Second World War, and Soviet resources were stretched by the effort to maintain political and economic control over eastern Europe, shaken by the Polish and Hungarian risings in 1956.[54] Despite its benefits to the Soviet Union's prestige, propaganda and deterrent posture, funding for the space programme was always well below what the design bureaux requested, and what was required to compete effectively with the United States during the 1960s.[55] The manned space programme was far from being a major priority for the Soviet state, which had many other pressing demands on the USSR's resources.

Moreover, the perception of Soviet capability that Khruschev had deliberately emphasised through the space programme produced not American defeatism or the acceptance of the USSR as an equal, but rather a vigorous and ultimately overwhelming counter-response from the United States and its European allies. The apprehension triggered by the Sputnik launch helped to rally public opinion in the NATO countries behind a programme of arms build-up and military improvements, accelerating competition rather than producing a new status quo manageable and acceptable for the Soviet Union. The dangers to the USSR represented by this energetic western response would subsequently encourage the Soviet leadership under Brezhnev to reduce the earlier emphasis on East–West competition and pursue a new policy of *détente* with the West instead. Once again, however, the space programme would be used to exemplify the new central Soviet propaganda theme.

President Kennedy's 1961 commitment of the United States to manned Moon landing by the end of the decade changed the dynamic of the space race. What, up to that point had been a series of sprints, now became more of a marathon, in which the USA could have hopes of prevailing. The Soviet space policy and programme were forced to react accordingly, although Khruschev's propaganda demands continued to be problematic for the orderly development of the Soviet programme. In October 1964, for example, the

USSR launched Voshkod into orbit, carrying a three-man crew. The mission was simply an effort to fly three men in space before the American two-man Gemini spacecraft became operational. The mission was designed to give the impression that the Soviet Union already had a three-man spacecraft, similar to the planned American Apollo lunar spacecraft, which was still three years away from its anticipated maiden flight. In reality, Voshkod was simply a slightly modified Vostok, in which the reserve parachute and ejector seat had been removed, along with the crew's spacesuits, in order to save weight. *Pravda* underlined the desired message with the headline 'Sorry Apollo!', and claimed that 'the so-called system of free enterprise is turning out to be powerless in competition with socialism in such a complex and modern area as space research'.[56] NASA was suitably impressed, but the mission had been scientifically meaningless, as well as technologically misleading, and had placed the lives of the crew at unnecessary additional risk.[57] The idea of a manned mission to the Moon never achieved the political backing that the American Apollo programme enjoyed. The series of space 'firsts' in the first half of the 1960s satisfied the Soviet leadership's requirements, while the central goal of achieving nuclear deterrence and military parity with the United States was being prioritised. Even the possibility of a joint US–Soviet lunar mission, proposed by President Kennedy at the United Nations in September 1963, was simply ignored by the USSR.[58] Perhaps appropriately, during the course of the Voshkod I mission, Khruschev was removed from power in the Soviet Union and replaced by a triumvirate dominated by Leonid Brezhnev.

The Soviet Union achieved huge propaganda benefits from its space launches between 1957 and 1964, despite the lack of a clear and coherent programmatic space plan. The most detailed western study of the programme argues that the Soviet space programme was never a major priority for the Soviet leadership, nor did Soviet leaders see it as a crucial instrument for achieving the USSR's ultimate ends. It was those working within the space programme who energetically promoted its many virtues for the Soviet Union, and their projects were 'grudgingly approved by the Communist Party and government, and then used as propaganda vehicles by Soviet leaders for selling the virtues of the socialist system'.[59] The propaganda advantages of the missions tended to be realised after, rather than before, the flights.

The successful American Moon landing of July 1969 led to the abandonment of the Soviet manned lunar programme. Rather than clearly come second to the Americans, the USSR chose to insist that it had no desire to go there at all. There was no great prestige or propaganda value seen in copying an American achievement. This decision reflected the way that the Soviet programme had been perceived in the previous few years, as an important dimension of Soviet military development, with spin-off activities that could be exploited for propaganda purposes. In some ways the decision to give up on the lunar project so easily is surprising given that it occurred at a time when the USSR was achieving parity rather than superiority over the USA in strategic nuclear weapons, and was pleased to see that parity, and the recognition as an equal that went with it, affirmed in the 1972 SALT I treaties with the United States. A desire to maintain a similar parity in space activities might therefore have been expected. In 1969, however, the Soviet Union had too much invested in the idea of leadership in space to surrender easily this image, at least in its own eyes, and it therefore sought to re-cast its programme in a way that suited its own purposes, and clearly differed from the American lunar focus, by developing manned space stations for operations in near-Earth space. Very quickly the USSR would realise that this too had significant political propaganda and prestige potentialities.

Chapter 3

The new frontier
US space policy 1957–72

We see another nation of great potentiality, militant and competitive, which has already made the first advances in the mastery of outer space. We cannot stand by and watch this nation make their mastery complete.

(Congressman John W. McCormack, House Majority Leader and Chairman of the Select Committee on Astronautics and Space, 15 April 1959)

The space race with the Soviet Union which the United States took up in 1957 was entirely the result of international politics, as the US endeavoured to contain the perceived damage to its self-perception as the world's leading scientific and industrial power, and as it responded to what it saw as a military as well as a political challenge. In attempting to respond as rapidly as possible, the USA faced a number of disadvantages. The United States entered the Cold War without the same tradition of official interest in long-range rocketry that had been typical of Germany and the Soviet Union during the inter-war period. Nevertheless, a number of factors helped the USA to develop rapidly a space capability once it committed its energies to doing so.

In contrast to the position in the Soviet Union, there had been no history of effective group research on space technology in the USA during the period between the two world wars. Nevertheless the US did benefit from the work of one remarkable individual, Robert Goddard. Goddard, a professor of physics at Clark University in Massachusetts, independently developed the concepts of liquid-fuelled and multi-stage rockets. During World War One he carried out research into military rockets for the US government.[1] Once the war was over, however, in a period of defence reductions and financial retrenchment, the US army lost interest in the military potential of rocketry, and Goddard returned to his civilian career. In 1919 he published the results of his pre-war research as a book, *A Method of Reaching Extreme Altitudes*, and in 1923 ended his military work and returned to his University.[2] In 1926 he carried out the world's first successful launch of a liquid-fuelled rocket and attracted research funding from private foundations. In the 1930s he moved his research to Roswell, New Mexico and by 1935 his rockets had passed the speed of sound.[3] Had these experiments attracted the interest and support of the military, as did the comparable activities of Korolev and von Braun in the USSR and Germany, the US would have been the world leader in rocketry by the end of the Second World War. This was not the case, however, and despite support from private foundations, Goddard's research was too poorly funded to compete with the German V-2 programme, although he solved most of the key questions in rocket engine design. Werner von Braun admitted that Goddard's work had been of enormous help to the German V-2 designers.[4]

In April 1940, Goddard's team launched the P-23 rocket. It was the most advanced rocket in the world outside Germany. The following month, with the Second World War almost a year old and the United States engaged in a defence build-up, Goddard met with representatives of the armed forces and offered the government total access to his rocket research. The military showed no interest in the potential of Goddard's work, however.[5] Within a year the situation changed dramatically. The Japanese attack on Pearl Harbor brought America into the war and Goddard returned to government work on the development of rockets for military purposes. He worked on this till his death in August 1945.

Although it was slow to perceive the military potential of ballistic missiles, the United States was impressed by the capabilities demonstrated by the Geman V-2 missiles in 1944–5. Well before the end of the Second World War, the USA had become determined to recruit as many as possible of the scientists and engineers who had worked on the V-2 project. In February 1945 they began compiling a list of the German rocket scientists the US most particularly wanted to bring to the United States.[6] In July 1945, Operation Overcast (later renamed Operation Paperclip), was initiated to locate and recruit the German rocket personnel. In addition to scientists, captured V-2 missiles were also brought to the US and launched at White Sands, New Mexico in a programme to evaluate their capabilities while studying the upper atmosphere. The last American V-2 in this programme was launched in 1952.

Alongside their new-found interest in ballistic missiles, the US armed forces were also beginning to evaluate the potential of orbiting satellites. As early as 1945 the US Navy sponsored research into the possibility of developing an artificial satellite.[7] The initial study concluded that such a satellite was feasible, and a contract to develop a launcher and satellite was given to North American Aviation, but lack of adequate funding and inter-service rivalry led to the 'Earth Satellite Vehicle' project being cancelled in 1948.[8] At the same time the US Army Air Force commissioned a RAND study which reported in May 1946 that a satellite launch might be achieved within five years. The report predicted that the initial satellite launch 'would inflame the imagination of mankind, and would probably produce repercussions comparable to the explosion of the atomic bomb'.[9] The report went on to outline the potential military uses of satellites, including weather forecasting, damage assessment, communications, navigation and weapons targeting.[10]

Satellite development was not accorded any special priority, however. Progress was held up by the same factors that hindered efforts to produce an effective launch-rocket; political indifference, military conservatism, inter-service rivalry and comparatively low defence spending in the late 1940s and early 1950s. At this time there was no effective US space programme as such. A number of organisations were carrying out rocket research on tiny budgets, but there was no overall coordination and a great deal of duplication of effort. The Air Force (Atlas), Army (Redstone) and Navy (Vanguard), all had separate rocket programmes and this fragmentation of effort significantly delayed the pace of development. The US was in any case not prioritising missile development, because unlike the USSR, the United States had a very large and effective long-range bomber force, capable of striking deep into the Soviet Union. US defence planners were therefore confident that for the forseeable future, a nuclear offensive could be carried out using aircraft, without the need to develop missiles. The missile programme therefore lacked the sense of urgency that characterised its Soviet equivalent. The sanguine US attitude was also due to a rather dismissive attitude towards Soviet technological and industrial capabilities, and an assumption that it would be a long time before the USSR began to pose a significant military-technological challenge to the USA.

In the mid-1950s however, the US effort began to accelerate. In March 1954 a USAF report recommended that the Air Force 'undertake the earliest possible completion and use of an efficient satellite reconnaissance vehicle' as a matter of 'vital strategic interest to the United States'. It recommended that the project, which it was estimated would take seven years to complete, should be conducted in the strictest secrecy.

Although the early military and scientific space research were almost indistinguishable, the decision to develop in great secrecy a military satellite alongside an open scientific space programme clearly marked the beginning of two avenues of development that would always be in tension thereafter. Despite a deliberate attempt at clear organisational partition after 1958, the two could never be kept completely separate, and as a result there were problems from the outset over the relative priorities, goals and public profiles of the civil and military space programmes.

Various factors had helped to gain approval for the satellite reconnaissance programme. The most important of these was the growing requirement for better strategic intelligence. This had been strongly advocated in the 1955 report of the Technology Capabilities Panel chaired by James Killian. The Killian Report, entitled *Meeting the Threat of Surprise Attack*, called for the development of advanced reconnaissance methods, including the use of satellites.

In May 1955, at a White House press briefing, the Eisenhower administration announced its intention to develop a 'scientific satellite' as part of the International Geophysical Year.[11] Because the development of a scientific satellite encroached on the existing and planned US military satellite and ballistic missile programmes, the National Security Council met in May 1955 to discuss guidelines for US participation in the project. The resulting directive, NSC 5520, *United States Scientific Satellite Programme*, decreed that the US satellite could not employ a launch vehicle currently intended for military purposes. While this decision was in part the result of a desire to enhance the peaceful image of the American space effort, it was primarily designed to protect the ballistic missile programme from diversion and disruption.

The decision to use a civilian programme to launch an American satellite during the IGY also reflected Eisenhower's desire to establish the legal legitimacy of satellite overflight of foreign territory, in order to allow a subsequent reconnaissance programme. Eisenhower felt that 'a satellite put up as part of the IGY program would strengthen the freedom of the skies policy and would be less likely to disturb Nikita Khruschev's sensibilities about overflight than one sponsored by the three military services'.[12] NSC 5520 made several references to the importance of establishing the freedom of space as a principle, a concern that would be central to US space policy over the subsequent decade.

NSC 5520 stressed the potential of reconnaissance satellites for enhancing US security, and declared that the intelligence applications justified an 'immediate program leading to a very small satellite in orbit around the earth'. The purpose of such a satellite would be to test the principles and technologies required to develop a subsequent large satellite capable of carrying surveillance equipment.[13] Given the uncertainties and fears of the early Cold War period, and the difficulties involved in determining the true military capabilities and intentions of a closed society like the Soviet Union, the focus on the need to establish both the capability and the legal right to monitor the USSR from orbit was an understandable American preoccupation.

The Eisenhower administration eventually decided to give the Navy Vanguard programme priority in the development of a satellite with a launch date set for 1958. Project Vanguard used the Viking sounding rocket. Given the constraints imposed by NSC 5520, Vanguard was an inevitable choice since it was the only one of the three service rockets that was based on a non-military launch vehicle.

If the purpose was to place a satellite in orbit before any other country, it was a poor decision. The US Army's Redstone Orbiter proposal, developed by a team of scientists under von Braun would have been a better choice, since it was a better launcher and further on in its development programme (as was demonstrated when a Redstone/Jupiter rocket eventually launched America's first satellite in January 1958). In any case, while the IGY satellite project was designed to be primarily a scientific enterprise, it was recognised from the start that certain military benefits might come from it. For example, NSC 5520 noted that scientific data on the upper atmosphere 'would find ready application in defense communication and missile research', and 'antimissile research will be aided by the experience gained in finding and tracking artificial satellites'.

The CIA commentary on NSC 5520 made a number of crucial observations, including that 'the psychological warfare value of launching the first earth satellite makes its prompt development of great interest to the intelligence community', that there was 'increasing evidence' that the USSR was planning to launch a satellite, that the Soviets had decided that the resources required were clearly justified for prestige or military purposes and that if the Soviets orbited a satellite before the United States did, 'there is no doubt but that their propaganda would capitalise on the theme of the scientific and industrial superiority of the communist system'.[14]

The United States was therefore stunned when the Soviet Union launched its own satellite first in October 1957. Public and congressional concern at the implications of the Soviet achievement created pressures for an accelerated and expanded US space effort. President Eisenhower had hoped for a fairly leisurely and orderly American entry into space, but instead found the United States gripped almost overnight by a national obsession to regain the scientific and technological lead apparently lost to the USSR. The rhetoric that followed the launch of Sputnik firmly established the idea of space as being central to national security. Eisenhower, despite his own reservations about the value of an energetic space programme, was forced to initiate an effort to compete effectively with the Soviets. 'Eisenhower's thinking throughout his presidency and in retirement was marked by the tension between his conviction that space exploration should not be the subject of international competition, and his recognition that it inevitably was'.[15]

As the RAND study had predicted, 'not since the explosion of the atomic bomb over Hiroshima had a technological event had such an immediate and far-reaching political fall-out'.[16] The New York Times declared that the Soviet rockets had launched a bombshell 'able to land on the front pages of every American newspaper'.[17] Americans, despite the messages of congratulations that were sent to the Soviet Union, felt that they had lost ground and 'face' in the competition with world communism. Senator Henry Jackson described the Soviet achievement as 'a devastating blow to the prestige of the United States as a scientific and technical world leader'.[18] The launch of Sputnik 'surprised the world, but shocked the United States'.[19] The unexpected achievement of Soviet science gave Americans 'both an inferiority complex and a heightened sense of vulnerability in what was then the most intense phase of the Cold War'.[20] Sputnik was seen as evidence of a vigorous missile development programme, particularly in regards to long-range nuclear missiles, and thus threatened the credibility of America's extended deterrence for its NATO allies.[21]

There followed a raft of congressional inquiries into the apparent complacency and inadequacy of the US missile and satellite programmes. These inquiries revealed that the United States in fact had an impressive programme of technological development under way in this field, but had failed to focus and concentrate its resources on a single programme, and had felt no particular sense of urgency given the availability of US long-range bombers, and the perception of the USSR as a technologically backward competitor.

In the aftermath of Sputnik, however, many Americans swung from one extreme to the other. Having seen no particular urgency in developing space technology, they now felt that the failure to do so revealed not simply a short-term technological or organisation shortfall, but rather a series of failings fundamental to the nature of American democracy and capitalism. Commentators criticised American capitalism, which was held to emphasise style over substance, so that instead of driving technical advances forward through its competitiveness, it was instead holding them back through its excesses and greed. The existing American educational system was seen as having failed to deliver, while even the effectiveness of the democratic system itself was questioned in regard to its ability to compete effectively with the single-minded ruthlessness of a dictatorship. The fear emerged that the satellite launch had indeed demonstrated the relative merits and advantages of the communist system, in comparison to its western rivals.[22] The National Security Council declared that if the Soviet Union maintained its lead in space exploration it would 'be able to use that superiority as a means of undermining the prestige and leadership of the United States and of threatening US security'.[23]

Science and technology became a major part of the political rhetoric of the period. The 1960 presidential election would be dominated by such language, with Democratic Senator John F. Kennedy promising to push America to a 'new frontier' of scientific achievement. Democrats in Congress claimed the existence of 'missile gap' produced by inadequate defence spending, which Eisenhower was forced to increase significantly in response to the crisis. Such criticisms played a major part in Eisenhower's decision to create the Presidential Science Advisory Committee (PSAC), chaired by James Killian. This 'summoning of scientists' was a measure taken largely to allay public fears.[24]

The drama of Sputnik triggered many changes for the US space effort. New organisational structures and procedures had to be formed to manage an expanded space programme. A complete review of US interests, priorities and goals in the exploration of space was also deemed necessary. In the initial government meetings following the Sputnik launch, Eisenhower reaffirmed the deliberate separation of the military and civilian space efforts, and emphasised the 'peaceful character' of the US programme.

Eisenhower initially resisted calls from the PSAC to create a new autonomous government agency capable of wielding the political and financial power necessary to drive the space programme forward in competition with the Soviet Union. The PSAC members were wary of the existing military dominance of space projects and advised that any new agency would need to be free from economic and scientific reliance on the US military. Eisenhower and his close advisers, however, felt that the military dimension of space activity must always take priority over civilian and that overall direction and control of the programme should therefore remain with the armed forces. He was wary about a further duplication of effort of the kind that had already hindered US progress, and did not wish to see American scientific talent diverted away from the defence programme.[25] However, his desire to maintain two clearly distinct programmes eventually led him to accept the arguments for the creation of a civilian space agency to parallel the military programme.

President Eisenhower vainly attempted to resist the clamour. He himself was not convinced that space was a crucial arena of political activity, and was opposed to the idea of joining a 'space race' with the Soviet Union. In signing the 1958 Space Act into law he made no mention of the USSR, and even after leaving office would continue to criticise his successor's vigorous space programme.[26] For Eisenhower, it was important that American policy should not simply be a reaction to whatever the Soviet Union was doing in space. This was particularly the case because at this time it was unclear as to the degree to which terrestrial international law was applicable beyond the Earth's atmosphere. The United States was only just beginning to gain a clearer picture of what its interests were in regards to space, and there was therefore a reluctance to undertake policy initiatives that might commit the United States before its interests were fully revealed. The Department of Defense cautioned that only once US interests and policy priorities had become clearer should it undertake major commitments or support initiatives in the development of a new international law applicable for outer space.[27]

On 1 October 1958, the National Aeronautics and Space Administration (NASA) came into existence, almost exactly a year after the launch of the first Sputnik. By creating a large civilian space agency, the administration hoped to send a clear signal that the United States did not want (at least overtly) to militarise space. The creation of NASA was accompanied by the passing of the National Defense Education Act, designed to expand and improve the American education system to allow it to compete more effectively with the Soviet Union, and regain its leading position as a leader of technological and scientific innovation. In addition to NASA, a National Aeronautics and

Space Council was created to advise the President, and Congress also set up a Civilian–Military Liason Committee to allow NASA and the Department of Defense to consult and advise one another. Neither worked well, and bitter interdepartmental battles between NASA and DOD became a feature of subsequent decades.

In 1958 the PSAC identified four drivers of the American space programme: the human urge to explore, the need to use space for military purposes to enhance US security, national prestige, and new opportunities for scientific discovery.[28] As with the Soviet Union, it was the second and third factors that would be the most compelling for the next two decades. The desire to regain and enhance national prestige would take centre stage in the short-term. Given the nature of the ideological competition between the two superpowers, prestige and national image were crucial not only in terms of how the United States perceived itself, but in terms of how the US was perceived by other countries. US statesman Bernard Baruch argued that 'we have been set back severely not only in matters of defence and security, but in the contest for the support and confidence of the peoples of the world'.[29] US foreign policy was driven by the need to win the hearts and minds of the population of America's allies, and the uncommitted nations of the 'Third World', the non-aligned states neither communist nor pro-American. There was also a need to impress the governments and peoples of the Soviet Union and its allies. In all cases, it was essential that the United States was able to project successfully an image of strength and leadership.

The 1958 Space Act declared that the United States was keen to explore space for 'peaceful purposes for the benefit of mankind', and allowed for 'cooperation by the United States with other nations and groups of nations'.[30] This declaration had a dual purpose. The first statement was designed to deflect attention away from the military dimension of US space research and reduce foreign concerns that the United States was seeking to militarise outer space. The second statement's purpose was to promote the image of the United States as a scientific leader that was willing to share the development of space with other nations, and which therefore clearly had no hidden agenda beyond space exploration for the general benefit of humanity. In this regard, it fitted in with other US policy initiatives designed to promote the image of the United States as a country eager to cooperate internationally in an open and transparent manner. The Marshall Plan, Atoms-for-Peace and the Peace Corps were all part of this general image-building approach, though all had other motivations as well, as did the space policy.

The apparent separation of civilian and military activities allowed the United States considerable flexibility. By having a largely transparent civilian-dominated programme, American public insecurity was alleviated, yet at the same time the US was able to continue its military programmes away from the glare of national and international scrutiny, and often successfully camouflaged behind actual or fictitious civilian space projects.

In fact, unknown to the American public, there were three, not two space programmes, white, blue and black. The white programme was the high profile civilian programme led by NASA. The blue programme was the classified military programme run by the Department of Defense. In addition, there was the 'black programme', the reconnaissance programme run by the intelligence agencies.

The apparent separation of the elements of the US space programme made it easier for the vast majority of the American political establishment to rally behind a substantial and energetic space programme. Liberals could support it as an alternative form of competition with the Soviet Union in an era when the dangers of nuclear war were very real, while conservatives saw the programme as developing military hardware and providing capabilities that would in the long run enhance the effectiveness of US armed forces.[31]

Concern over the poor management and progress of the satellite reconnaissance programme led to the creation of the highly classified National Reconnaissance Office in 1960. Eisenhower had directed the Central Intelligence Agency to develop its own reconnaissance satellite, and the establishment of the NRO stemmed from the growing CIA involvement in the satellite reconnaissance effort. The CIA project, codenamed Corona, was funded partly from CIA funds and partly from the US Air Force's Discoverer programme, which acted as its cover. Officially Discoverer capsules would take mice and monkeys into orbit as part of biomedical research. In reality, the purpose of the Discoverer programme was to orbit photo-reconnaissance satellites over the Soviet Union.[32] In this regard, however negative the public relations impact of Sputnik had been for the United States, it did at least produce one huge benefit. The repetitive orbits of Sputnik produced no protests from the states over whose territories it passed, and thereby 'with the lack of worldwide objection to overflight, Sputnik I literally wrote overflight into international law'.[33]

Sputnik had profound consequences, both for the American space programme and for US foreign and defence policy more generally. The extreme nature of the American response to Sputnik and its impact on subsequent policy is all the more remarkable given that the launch ought not to have come as a surprise. The Soviet Union had not shrouded its programme in secrecy, on the contrary it had announced its plans to launch a satellite on numerous occasions, and subsequently insisted that it would achieve this feat during the 1957–8 International Geophysical Year. America was taken by surprise only because it had chosen to dismiss the Soviet announcements as wishful thinking and baseless propaganda.

Nevertheless, it is not the case that the subsequent US space programme was no more than a reaction to the Soviet achievement. Rather it served to rapidly consolidate and accelerate a policy trajectory that had already taken shape. The United States, like the Soviet Union, had announced that it would launch a satellite before the end of the IGY, and like the USSR, believed that this would provide a major boost to national prestige.[34] The Eisenhower administration was also committed to the development of reconnaissance satellites as a crucial requirement for US national security in an age of nuclear deterrence and rapidly improving anti-aircraft defences. The US was also developing long-range missiles as a future addition to the aircraft-delivered nuclear deterrent. Thus though the US response to Sputnik was dramatic and somewhat *ad hoc*, it also embraced policies that had been under development for several years already.

The technological progress represented by the Soviet satellite launch was associated by the American public, and much of the élite, with a clear threat to US national security. To some extent this was based on an irrational fear of the communist threat, compounded by the sudden sense of inferiority. The accelerated space programme was to a significant extent 'an emotional reaction to the Sputnik blitz'.[35] These concerns were fuelled by near-hysterical and often highly exaggerated press coverage of Sputnik and its aftermath. *Newsweek Magazine* for example asked whether 'the crushers of Hungary could be trusted with this new kind of satellite whose implications no man could measure'.[36] In similar vein, after the failure of the Vanguard rocket launch in 1957, the *Washington Post* announced that the United States was 'in the greatest danger in its history' with 'America's long-term prospect one of cataclysmic peril in the face of rocketing Soviet military might'.[37]

This rhetoric was important in placing space firmly within the national security agenda. Prestige alone was important, but might not have been enough to trigger the funding flow for space technology that subsequently materialised. But the reality was that 'American space activity, whether military, scientific, civil or commercial, emerged originally from within the driving issue of national security'.[38]

In practice, the United States was as far advanced as the Soviet Union in terms of its conception of the potential value of space, and was not dramatically behind the USSR in terms of the associated research and development programme. After Sputnik, however, the funding floodgates opened, allowing a rapid acceleration in development. At the same time the consolidation of the previously fragmented missile development programmes gave the US programme a coherence and momentum that it had previously lacked. As Congress was creating NASA, the Department of Defense was creating the Advanced Research Projects Agency and accelerating the development of the Minuteman ICBM. Nevertheless, despite the post-Sputnik funding frenzy, not all proposed space-related projects were accepted. A number were dismissed because they were seen as being too provocative to the USSR or to international opinion more generally. These included the MIDAS early-warning satellite, Advent communications satellite and the Manned Orbiting Laboratory.

Some commentators such as Etzioni have suggested that the US badly overreacted to Sputnik, and that the satellite launch had not in fact led other countries to reassess their view of the relative positions of the United States and the Soviet Union.[39] At the élite level this may have been true, although public opinion surveys in key countries suggested that opinion had been significantly affected, at least in the short term. In any event, while such a conclusion may be possible with hindsight, it was not obvious at the time, and it would have been dangerously complacent to make such an assumption at the time, rather than reacting vigorously and effectively. In the final analysis, 'the impetus for US space exploration was linked, not to any compelling scientific rationale, but to the political exigencies of the Cold War'.[40] Notwithstanding the excitement generated by the Kennedy presidency, the late 1950s and early 1960s were a period of difficulty for the United States, with overseas crises, rising racial tension and unemployment, and with a quarter of the American population living in poverty.[41] American superiority over the Soviet Union was not something that could be simply taken for granted, nor could it be automatically assumed that other nations would automatically accept the superiority of the American social, political and economic system over that of the Soviet Union. It is noticeable that opinion polls taken in the United States during the 1960s showed that while 'public approval of the US space programme generally jumped after a successful Soviet effort, the approval rating was almost unaffected by American achievements'.[42] It was clear that the American public saw the programme in the same terms as Presidents Kennedy and Johnson.

Sputnik did not so much change US policy as act as a catalyst to an accelerated and more broadly-based implementation of existing policy. Much of the effort that followed was of little value in terms of scientific research, but was crucially important in terms of self-perception and relative superiority in the superpower competition. Eisenhower's science adviser reminded the President in 1960 that 'at present the most compelling reason for our effort has been the international political situation, which demands that we demonstrate our technological capabilities if we are to maintain our position of leadership'.[43] Sputnik did shake the US out of a period of relative complacency, and after the initial scare, subsequent US policy developed in a coherent and effective manner.

A mere twelve months after its creation NASA published an optimistic ten year plan. Its ambition was considerable, given the agency's infancy and its very limited achievements up to that point. The plan proposed 'manned orbital missions, the launch of robot probes to Mars and Venus, the landing of unmanned surveyor craft on the Moon, and the establishment of a permanent near-Earth space station'.[44] Ironically, while a manned spaceflight around the Moon was seen as possible during the 1960s, no manned landing was envisaged before the 1970s. While unmanned missions might be more valuable from the national security perspective, manned missions were seen as central to winning the propaganda contest with the Soviet Union.

On being elected President Kennedy immediately appointed a committee under Jerome Wiesner to review national space policy. The Wiesner committee reported two weeks before Kennedy's inauguration and insisted that 'manned space flight was an expensive and risky public relations gimmick'.[45] Surprisingly, given all the rhetoric about the space race that had marked the previous three years, Kennedy's reference to space in his inaugural address called for the two superpowers to 'explore the stars together'. Kennedy would return to this idea later in his presidency, and encouraged his advisors to identify specific proposals for superpower space cooperation, but did not pursue it in the months after his election. Instead, the competition with the USSR was emphasised, and in 1961 Kennedy committed the United States to the goal of a manned lunar landing, a mere four months after becoming President. Addressing a special session of Congress on 'Urgent National Needs', Kennedy announced that 'I believe this nation should commit itself to achieving the goal, before this decade is out, of landing a man on the Moon and returning him safely to the earth'.[46]

The prestige rationale was made very evident in Kennedy's remarks. 'No single space project in this period will be more impressive to mankind, or more important, or so difficult and expensive to accomplish ... In a very real sense, it will not be one man going to the Moon, it will be an entire nation'.[47] Even so, the commitment represented a major gamble. The applause for Kennedy's speech was unenthusiastic, and his own Presidential Transition Team had warned him that the United States was unlikely to win a space race with the Soviet Union.[48] Polls taken immediately after Kennnedy's speech showed 58 per cent of the American public opposed to the Moon programme.[49] Nevertheless, there had been a logic in Kennedy's ambitious space goal. Since the USSR was ahead in the space race, which was being conducted as a sprint, the US needed to turn it into a marathon, in which it would have time to catch up and overtake its superpower rival. NASA adviser Robert Gilruth had told Kennedy that 'you've got to pick a job that's so difficult, that's so new, that they'll have to start from scratch, so they can't just take their old rocket and put another gimmick on it and do something we can't do'.[50]

In some ways the decision by Kennedy to offer such a dramatic challenge was surprising. In the months leading up to the speech he had not shown any particular enthusiasm for an energised national space programme. After delivering his critique of the space programme, Jerome Wiesner had been appointed the President's Special Assistant for Science and Technology, with a remit covering all space matters, and Kennedy had left the membership of the President's Science Advisory Council unchanged from the Eisenhower era. In his inauguration speech, far from committing himself to an accelerated national space programme, he had instead suggested cooperation in space research between the USA and USSR, suggesting that 'both sides seek to invoke the wonders of science instead of its terrors. Together, let us explore the stars'.[51] Perhaps because of the spiralling costs of the Apollo programme, and perhaps in response to the sense of potential *détente* that followed the successful conclusion of the nuclear Partial Test Ban Treaty in 1963, Kennedy returned to the possibility of superpower space cooperation in the speech given to the UN General two months before his death in November 1963. He proposed that the USA and USSR should contemplate a joint lunar mission, both to share the expense and challenges involved, and because then the issue of potential territorial and sovereignty claims to the Moon would not become a problem.[52]

A number of incidents in both the space programme and the international political environment had brought about Kennedy's change of attitude, because they were felt to have badly damaged the self-confidence and international standing of the United States, and because they reflected badly on his own administration. On 25 March 1961, an Atlas ICBM, carrying an unmanned Mercury capsule to orbit, exploded at 35,000 feet. It was the latest in a succession of similar setbacks, the United States having experienced 16 such flight failures in 1959 and 1960.[53] Eighteen days later, Soviet Air Force Major Yuri Gagarin became the first human being to fly in space, triggering a new round of media and political comment about the relative technological capabilities of the two superpowers.[54]

Theodore Sorenson, Kennedy's speechwriter and biographer, felt that the Gagarin flight, more than anything else, triggered Kennedy's commitment to establish US superiority in space technology.[55] But Kennedy was reluctant to challenge the Soviets directly in an area where they seemed to have a clear lead, declaring at the post-Gagarin presidential press conference that the United States would not attempt to match directly Soviet achievements in space, but would instead choose 'other areas where we can be first'.[56] The Gagarin flight was in some ways a worse blow to US prestige than the Sputnik launch had been, since by 1961 the American people and US allies abroad were aware that the US had been engaged in a competition to launch an astronaut before the Soviet Union did.[57]

At this point Werner von Braun directly contacted vice-President Johnson, suggesting that the United States would have an excellent chance of beating the Soviets in a race to land men on the Moon.[58] Johnson was a receptive audience, having declared while still a Senator that Sputnik posed an even greater threat to the United States than the Japanese attack on Pearl Harbor had, and that 'control of space means control of the world'.[59] Prior to becoming Kennedy's vice-President, Johnson had been chair of the Senate Space Committee, and on taking office Kennedy made him chair of the Presidential Space Council. By giving this role to a politician who had used Sputnik to such political effect in attacking the Eisenhower administration, 'Kennedy had tipped the scales in the direction of an aggressive effort in space'.[60] Kennedy had already decided that it was now essential for the United States to be seen to match and surpass the Soviet achievements in space.[61] On 21 April 1961 Kennedy sent a memo to Johnson asking him to identify an objective, such as orbiting a manned laboratory or landing astronauts on the Moon, which would be politically dramatic and where the United States was sure it could win in a competition with the USSR.[62]

The final, and perhaps decisive factor in changing Kennedy's commitment was the disaster of the Bay of Pigs invasion. The US-backed attempt by exiled Cubans to invade Cuba and overthrow the Castro government was a catastrophic failure. Although devised and planned by the Eisenhower administration, the blame for the failure fell on the incumbent President Kennedy.[63] Coming only 15 days after the Gagarin flight, the administration had to cope with two major foreign policy set-backs in a fortnight. In different ways, the Soviet spaceflight and the failure to overthrow Castro both suggested the resilience and effectiveness of communist regimes. Both set-backs seriously damaged the prestige of the United States, and Soviet leader Khruschev suggested that Kennedy lacked strength of character.[64] Kennedy responded with a press conference statement that 'if we can get to the Moon before the Russians, then we should',[65] and the following month made his address to Congress committing the United States to the lunar landing.

The lunar goal was a deliberate attempt to demonstrate the technological and organisational superiority of the United States *vis-à-vis* the Soviet Union, and by extension, of the superiority of democracy and capitalism over communism and the centrally planned economy.[66] The Moon programme was well suited to the goal of demonstrating this, and of allowing the United States to catch up with and surpass Soviet achievements in space, because it was technologically demanding, extremely expensive and would inevitably take either country several years to achieve. Kennedy had also already been informed that it was unlikely that the United States would succeed in beating the Soviet Union with a suggested alternative project, a manned orbiting space station.[67] Kennedy told Wiesner that political considerations were at the root of his decision to initiate the Apollo moon programme.[68] Both NASA and the Pentagon agreed with this perspective. A joint report by the two organisations to the vice-President in May 1961 insisted that the prestige consideration was so important that it fully justified the space programme, 'even though the scientific, commercial or military value of the undertaking may by ordinary standards be marginal or economically unjustified'.[69]

That the Apollo programme was a response to Gagarin's flight and the Bay of Pigs fiasco is also suggested by the fact that just prior to Gagarin's flight NASA Director James Webb had met with Kennedy in March 1961 to request funding for the Apollo spacecraft, only to have the request refused and put on indefinite hold.[70] But the Apollo decision made sense in the wider context of Kennedy's overall approach to the confrontation with the USSR. 'In contrast to Eisenhower, Kennedy held that the struggle with the Soviet Union had to be waged in every category of power and in every part of the world'.[71] For Kennedy, 'the stakes in the competition with international communism were too high to play with a weak hand. It was time to take the (Space) Race to a higher level'.[72] It was a view that NASA and the Department of Defense were happy to endorse. Their May 1961 joint report to the White House argued that 'it is man, not merely machines in space, that captures the imagination of the world … dramatic achievements in space symbolise the technical power and organising capacity of a nation … the non-military, non-commercial, non-scientific, but 'civilian' projects such as lunar and planetary exploration are, in this sense, part of the battle along the fluid front of the Cold War'.[73]

The United States adopted a safety-conscious approach in developing its manned space programme, conscious that the launches would deliberately take place in the full glare of the international media. Two sub-orbital manned flights were made, with the spacecraft entering space on a ballistic trajectory, but returning without going into orbit, before John Glenn became the first American to emulate Gagarin by entering Earth orbit. The mission took place on 20 February 1962, nearly a year after Gagarin's flight. However, having proved that the technology worked, NASA carried out two more Mercury one-man missions, with Gordon Cooper's flight in May 1963 lasting 34 hours.

The one-man Mercury spacecraft was followed by the two-man Gemini. A lunar mission would take over a week, involve rendezvous and separation in space, and require delicate manoeuvring. The Gemini series of spacecraft were used to develop and master these techniques. The American approach to the Moon landing was an extremely cautious one, but their methodical development of the Gemini series gave them the experience and technological development that were required to make the lunar landing a success. There were two unmanned Gemini launches in 1964 and 1965, before Grissom and Young were launched on the first manned Gemini in March 1965. Subsequent missions steadily developed required procedures. Gemini 4 saw the first American spacewalk, Gemini 6 rendezvoused in orbit with Gemini 7, whose crew completed a two-week long duration flight. Gemini 8 saw the first successful docking of spacecraft in orbit. The series ended with Gemini 12.

The pattern of steady incremental progress was designed to continue in 1967 with the introduction of the three-man Apollo spacecraft, the vehicle that would actually carry out the mission to the Moon. However, on 27 January 1967, during a static ground test of the capsule, a fire broke out killing the three astronauts inside. Manned launches were halted until the problems with the spacecraft were rectified, and it was a year and a half before they were resumed. Once again, a series of unmanned Apollo launches preceded the first manned launch, Apollo 7. It was followed by Apollo 8, whose crew became the first humans to leave Earth orbit, carrying out a successful circumnavigation of the Moon in December 1968. Apollo

10 followed in May 1969, carrying out all the elements of the lunar mission except the landing itself, which was achieved successfully by Armstrong, Aldrin and Collins, the crew of Apollo 11 in July 1969. With Apollo 11, NASA achieved Kennedy's objective of a successful manned lunar landing and safe return to Earth, with five months to spare of the deadline the President had set in 1961.

Kennedy's political decision to prioritise a vigorous lunar programme would have significant implications for the long-term development of US space capabilities. The overriding priority given to the goal of putting men on the Moon before the end of the 1960s interfered with NASA's ability to develop across-the-board space capabilities in a measured manner, and use them for a variety of purposes in Earth orbit and beyond. The hardware developed by NASA was specifically designed for the lunar missions, and in the less pro-space political environment that followed the election of President Richard Nixon in 1968, the NASA technology would prove difficult to adapt to the kind of programme Nixon was willing to support.[74] After the Apollo 17 mission in 1972 it was employed only in adapted form for the Skylab missions, and for the Apollo-Soyuz Test Project in 1975, the latter a purely political symbol of *détente* between the superpowers.

In his address to Congress in 1961, President Kennedy had claimed of the lunar landing that 'no single space project in this period will be...more important for the long-term exploration of space'.[75] However, although Kennedy had given NASA a focus for the 1960s, he had not committed the United States to NASA's long-term vision, only to part of it.[76] This was important, for as NASA's Associate Director for Manned Spaceflight pointed out shortly after Kennedy's address, 'many people seem to believe that a landing on the Moon ahead of the Soviets is our paramount objective. But this is not so. The principal goal is to make America first in space generally'.[77] But Kennedy made no space commitments beyond the Moon landing, and in emphasising that particular objective he distorted and eliminated many of the steps NASA had built into its long-range plan.[78] The major example of this was the goal of developing a space station. This had been NASA's primary objective in 1961, but was not finally approved until the advent of the Reagan presidency, a gap of 23 years, having been rejected by every intervening administration.

NASA subsequently struggled to cope with the 'lack of a widely understood purpose, direction, and time scale for the manned space programme'.[79] At each stage of the US programme, there was an absence of consensus on what exactly the United States wanted from its space programme, bar the short-term goals that suited the political community at that particular time. Nor should NASA be seen simply as a victim of politics. Its projects flourished during periods of intense antagonism and tension with the Soviet Union in the early 1960s and 1980s, and the agency had been happy enough to emphasise the competition and its importance to America's image when it felt that this would help its funding requests. The problem for NASA after 1969 was that it had, for its own reasons, deliberately played up the idea of the United States being in a 'race' with the Soviet Union in space. In 1969 the USA won that race, and as Smith points out, 'once a race is won, only a fanatic keeps running'.[80]

It was also the case that Kennedy had initiated two space races. The first was an attempt to regain prestige by surpassing the USSR in manned space exploration. The second was the race to be pre-eminent in the military exploitation of space. Everett Dolman goes so far as to suggest that Kennedy and Johnson were the original practitioners of 'astropolitik' defined as 'a determinist political theory that manipulates the relationship between state power and outer-space control for the purpose of extending the dominance of a single state over the whole of the Earth'.[81] This form of global geopolitics is simply an extension of terrestrial realism, and early American space policy certainly seems to have been driven by a classical Cold War realism.

However, while there is no doubt that Kennedy viewed the space programme purely as a propaganda tool, he appears to have remained genuinely undecided about the best way to use it in terms of shaping the relationship with the Soviet Union and international perceptions of the two superpowers. While he frequently advocated the programme as a crucial element in the global competition with communism, he also frequently returned to the idea of using space as an arena to demonstrate the possibilities of US-Soviet cooperation and even rapprochement. The most dramatic example of this sentiment was his 1963 speech at the United Nations where he asked 'whether the scientists and astronauts of the two countries – indeed of all the world – cannot work together in the conquest of space, sending some day in this decade to the Moon, not the representatives of a single nation, but the representatives of all our countries'.[82]

Lyndon Johnson, who succeeded Kennedy as President after the latter's assassination in November 1963, shared his predecessor's concern with the importance of space policy and the perceptions of national capability that went with it. For Johnson, 'failure to master space means being second best in the crucial arena of our Cold War world. In the eyes of the world, first in space means first, period. Second in space is second in everything'.[83] Johnson's perceptions would ensure that the effort to surpass clearly the Soviet Union in space would be sustained throughout the 1960s. By the middle of the decade, however, it was clear that the United States had overtaken its rival, and references to the space race theme began to diminish significantly in government statements on the programme.

Unlike President Johnson, former President Eisenhower, in contrast, never became reconciled to the value of the manned space programme, criticising Kennedy in retirement by declaring firmly that 'anybody who spends $40 billion in a race to the Moon for national prestige is nuts'.[84] But while the space race was about prestige, it was more than just a simple desire to swagger on the world stage. It was an important surrogate for war in the contest for global domination that the Cold War represented. It was this factor that assured the space programme of such heavy funding for so long. 'Engaged in a broad contest over the ideologies and allegiances of the non-aligned nations of the world, space exploration enjoyed for more than a decade a treasured place in the pantheon of American public policy initiatives'.[85]

The successful manned lunar landing and safe return of Armstrong and Aldrin in July 1969 represented the achievement of a number of important goals for the United States, both in terms of the Cold War military/technological competition and the prestige and propaganda objectives pursued since the shock of the Sputnik launch. Sputnik had produced profound alarm in the United States and the successful conclusion of the Moon landing set the final seal on the process of recovery and response to the perceived Soviet challenge. The lunar triumph firmly established the international perception that the United States was the technological superior of the Soviet Union, a perception that would not be significantly influenced by the success of the Soviet space station programme in the following decade. The American triumph may have even encouraged a degree of hubris, with some Americans coming to believe that 'winning the Moon race showed the merits of free enterprise, the American way of life, and perhaps, Christianity'.[86]

However, the space achievements proved only a partial anaesthetic to the major problems dividing Americans at the end of the 1960s, the war in Vietnam, racial division, poverty, political and social violence, growing economic difficulties and so on. The Moon landing generated an enormous sense of self-confidence in the United States, particularly reinforcing the belief that the US could find technological solutions, such as the nuclear weapon, and the Moon landing, to problems facing it in the realm of international politics. For former NASA flight director Chris Kraft, Apollo was 'the best thing America did in the twentieth century'.[87] Certainly, the technological 'can do' belief system it confirmed would have important repercussions for American foreign policy and international relations in the decade that followed.

Overall, the United States clearly recovered well from the shock of Sputnik, which ultimately led to a US programme that overshadowed the Soviet achievements. A 1971 assessment of the programme concluded that it had made a highly positive contribution to American diplomacy and that 'our nation's strength and national security have been augmented, and new channels have been provided for significant enhancement of the partnership we seek with friendly nations and for the successful negotiation of agreements with the Soviet Union'.[88]

But the successful lunar programme, while politically successful in re-establishing international perceptions of American technological and therefore political superiority, left unanswered other crucial political questions.

During the remainder of the Cold War, the United States would continue to wrestle with the need to contain the dangers posed by an arms race in space, and the continuing question of how to exploit the non-territorial realm of space for various purposes within the context of an international system composed of territorial states.

Chapter 4

International cooperation in space

We do hope and we do pray that the time will come when all men of all nations will join together to explore space together, and walk side by side toward peace.

(President Lyndon Johnson, telephone conversation with Gemini astronauts McDivitt and White, 7 June 1965)

The drama of the space race between the superpowers in the 1950s and 1960s, together with the overwhelming use of space for military purposes, easily creates an impression of space as a realm of conflict and danger. However, the reality of the space age has been that space activities have been characterised by an enormous amount of international cooperation. This can be seen in the programmes of individual states, in various multilateral international programmes, in the dramatic cooperation of the western European countries, and in the work of international organisations, particularly those that operate under the structure of the United Nations. The United States recognised this even during the Star Wars tension during the mid-1980s, the Office of Technology Assessment noting that, '[s]pace is by nature and treaty an international realm about which cooperation between nations on some level is essential, if only to avoid potential conflict over its resources'.[1]

However, it was the Soviet Union that first seriously exploited this feature for political purposes. For the Soviet Union under Khruschev, the space programme had been a weapon of the Cold War competition between the superpowers, but once the USSR had lost the race to the Moon, his successors turned the programme into an instrument for the promotion of *détente* and international cooperation. As with the earlier phase, symbolism was all-important, and propaganda was used to ensure that the message the USSR was attempting to convey was clearly understood. Thus, although it took a rather more benign form, propaganda continued to be at the heart of the space programme as the Soviet Union once again sought to exploit it for political purposes and the advantages it could yield in foreign policy.

Two clear themes are notable in the post-1969 Soviet programme. One was international cooperation, but the second was the steady increase in the exploitation of space for military purposes, most notably for reconnaissance and early warning. These two themes are reflected in the tone and content of Soviet space-related propaganda in this period. On the one hand, Soviet space achievements were eulogised as reflecting humanistic principles. On the other hand, in order to deflect attention away from the military aspects of the Soviet programme, Soviet propaganda continuously argued that the United States was seeking to 'militarise' space.

Within this overall pattern, there was a shift in the sub-theme present under Khruschev. Khruschev had sought to identify the successes of the programme with himself. Under his successors, the Soviet propaganda machine consistently sought to identify Soviet achievements with the Communist Party of the Soviet Union, and with the Soviet Government as a whole, emphasising that both, but particularly the Party, were the source of the Soviet successes. In a state that was an uneasy combination of many nationalities besides the dominant Russians, space achievements were used to solidify Russian support for the Communist Party, and to encourage an incipient Soviet nationalism, by evoking the pride of the population in a successful and international prestigious endeavour. Soviet spokesmen confidently asserted that 'the Communist Party and the Soviet government created the necessary economic, social, scientific and technical conditions for the development of cosmonautics, and as a result of this the world's first socialist state opened the road to the stars for mankind'.[2]

The space programme continued to benefit the Soviet Union's image as a technologically dynamic industrial superpower, and a leader in the most advanced fields of science and technology. In addition, the international missions that were to become a feature of the 1970s and 1980s presented a positive image of the USSR. The inclusion of foreign cosmonauts in Soviet manned space missions presented the USSR as a country open to cooperation, with nothing to hide in its space programme, and happy to share the prestige of space exploration and its tangible benefits with other countries. The political importance attached to this cooperation was stressed by President Brezhnev in a speech to the 26th Congress of the Soviet Communist Party in 1981, when he insisted that 'the cosmonauts of the fraternal countries are working not only for science and for the national economy, they are also carrying out a political mission of immense importance'.[3]

The Soviet Union continued to use the successes of the space programme for political advantage in a number of ways. One such was the use of cosmonauts as ambassadors-at-large, using their fame and recognisability to promote a positive image of the USSR, and to give encouragement and public support to communist governments and parties around the world. Their fame was deliberately linked with the particular political ideas and policy lines being advocated by Moscow and, more broadly, with historic communist traditions.

In October 1977, for example, Valentina Tereshkova, who had been the first woman to fly in space, visited London at the invitation of the Communist Party of Great Britain, to participate in celebrations to commemorate the sixtieth anniversary of the Bolshevik revolution in Russia. Tereshkova was by this time a member of the Central Committee of the Soviet Communist Party, and Chair of the Soviet Women's Committee. After her 1963 spaceflight she had become a roving goodwill ambassador for the USSR, particularly among international women's organisations.

German Titov, who had become the second person to fly in space, was similarly active at the head of the Soviet-Vietnamese Friendship Society, an important role in the context of the Cold War conflict in Vietnam and the desire of the Soviet Union to expand its influence in South-east Asia, both as a counter to the massive American military presence there, and because of the dangerously adversarial relationship between the USSR and China.

The Soviet Union, like the United States, used its remote sensing satellites to build cooperative links with other countries. From 1966 onwards the USSR began to share imagery from its meteorological satellites through the World Meteorological Organisation. Imagery obtained from Salyut missions was also made available to developing countries, some of which were Soviet allies, and some that were not, for example Cuba, Vietnam, Morocco and Angola.[4]

A more dramatic example of the symbolic use of the space programme for political purposes was the series of joint flights with cosmonauts from communist states that began in 1978. These had been made possible by the development of orbiting space stations by the Soviet Union in the early 1970s. After losing the race to the Moon in 1969, the USSR re-oriented its manned space programme towards the goal of placing space stations in low-Earth orbit. In October 1969, the USSR launched three Soyuz spacecraft and manoeuvred them simultaneously in orbit to within a few hundred metres of each other. A year later a two-man Soyuz crew carried out an 18-day mission in orbit. These flights were crucial in demonstrating the capabilities needed to link with a station in orbit, and the resilience of human beings on long-duration space missions, which would be required of a space station crew.

Nor had the Soviet Union completely given up on the Moon. In the autumn of 1970, Luna 16 successfully soft-landed on the Moon, took on board a sample of lunar soil and blasted off to a safe recovery on Earth. It was an impressive technical feat and the Soviet Union sought to gain propaganda mileage from it by stressing that it had cost vastly less than the American Project Apollo manned lunar programme, and had put no human lives at risk in bringing back much the same knowledge as Apollo 11 had done. Two months later, the USSR landed another spacecraft which placed Lunarkhod 1 on the Moon's surface. This roving wheeled research vehicle (one giant leap for robotkind) spent the next 11 months manoeuvring over the lunar surface. The Soviet Union also demonstrated its technological prowess by sending several probes deeper into the solar system, notably Venera 7, which successfully penetrated the intensely hostile acidic atmosphere of Venus and transmitted signals from the equally unpleasant surface (temperatures of 475 degrees centigrade and a surface atmospheric pressure 90 times that on Earth).

The launch of the key manned missions was typically linked to political considerations. In April 1971, the USSR chose the tenth anniversary of Yuri Gagarin's flight to launch its first space station into orbit. The Salyut stations were much larger than the Vostok and Soyuz spacecraft, making long-duration missions possible. Soyuz 10 was launched to dock with Salyut, but the crew were unable to gain entry and returned to Earth. It was therefore the cosmonauts of Soyuz 11, Dobrovolski, Volkov and Patseyev, that were the first Salyut crew. In contrast to the secrecy typical of the early Soviet spaceflights, the now confident USSR beamed daily television images of the crew's activities to the Soviet population. The three cosmonauts became instantly recognisable national heroes. This familiarity made it all the more shocking when a depressurisation accident killed the crew during their re-entry through the Earth's atmosphere. Western diplomats reported that the reaction of the Soviet people was comparable only to the national trauma suffered by Americans as a result of the assassination of President Kennedy in 1963. Soviet leader Brezhnev wept openly at the state funeral of the cosmonauts. As with the earlier Voshkod, the crew of Soyuz were not wearing pressure suits, to save weight. The suits would have saved their lives had they been wearing them. Soyuz never flew with a three-man crew again, and it was two years before another manned flight by the USSR was launched.

Salyut 1 tumbled from orbit in 1971. The follow-on stations incorporated modifications and improvements, including a second docking port on Salyut 6. The second port was needed so that unmanned resupply spacecraft could visit the station while the crew's spacecraft was docked at the first docking port. However, once unloaded, these supply craft were undocked and allowed to burn up in the Earth's atmosphere, leaving the docking port available. The addition of the second docking port had a revolutionary impact, making it possible to carry out long-duration missions.[5]

This feature also made it possible for shorter-duration visiting crews to be sent up to visit the long-term crews on the Salyut station, and it was this that gave birth to the Intercosmos programme, where one of the two visiting cosmonauts would be a 'guest' from outside the Soviet Union. The Intercosmos flights were invariably accompanied by frequent expressions of mutual friendship, solidarity and support for common policies. For example, in March 1978, Czech cosmonaut Vladimir Remek flew on the Soyuz 28/ Salyut 6 mission, becoming the first space traveller who was not an American or Soviet citizen.[6] In a speech marking the flight in May 1978, Joseph Lenart, the leader of the Czechoslovak Communist Party, claimed that the flight had demonstrated the communist countries' desire to use science for national economic development and wellbeing, whereas the United States military-industrial-complex sought only to expand the arms race.

Cosmonauts in orbit routinely sent politically inspired congratulatory messages to the Soviet leadership. For example the Salyut 5 crew assured President Brezhnev that 'we would like to dedicate this flight to the forthcoming sixtieth anniversary of the great October Socialist Revolution'. An article on 'Cosmonautics and social development' in the November 1977 issue of *International Affairs* (Moscow) in many ways sums up the key themes in Soviet propaganda relating the space programme to Soviet political goals. The article is a comprehensive statement of the Soviet attitude toward space exploration placed within a political context. Key themes include the idea that the space programme of the socialist states has the profoundly humane task of making outer space and space technology serve man, whereas western programmes are driven by military considerations and the desire of capitalist companies to make 'immense' profits.[7] The link between the visible successes of the space programme and the virtues of the communist political and economic system, a theme present since 1957, is also prominent. Thus, 'the development of theoretical and practical cosmonautics in the Soviet Union and the constant fulfilment of most sophisticated space programmes are striking evidence of the advantages and potential of socialism'.[8] The central role of the party in making these achievements possible is also emphasised.[9]

The inauguration of the Intercosmos programme began in December 1976 when foreign cosmonauts began flight training. The Soviet Union saw the programme both as an important symbolic effort that lent itself to positive propaganda, and also as an integrating force for consolidating the unity of the Soviet bloc. Guest cosmonauts were 'flight engineers' and 'mission specialists'. By the end of the programme cosmonauts from all the Warsaw Pact states as well as the other communist countries in the world had flown Soyuz/Salyut missions.

The identity of the countries selected, and the order in which their cosmonauts flew in space was highly politicised. The first such flight was Soyuz 28, taking Vladimir Remek in 1978. Remek as a Czechoslovak was deliberately chosen first because 1978 was the tenth anniversary of the controversial Soviet invasion of Czechslovakia, which had brought the 'Prague Spring' of Alexander Dubcek's liberal communist regime to an end in 1968. The Soyuz flight therefore sought to emphasise the closeness of Soviet–Czechoslovak cooperation and the USSR's recognition of Czechoslovakia as a sovereign equal of the USSR within the Warsaw alliance. On Remek's return to Earth, Czechoslovak political leaders vied with each other in emphasising the political significance of the mission. The President, Gustav Husak declared that it was 'above all, a manifestation of the internationalist policy of the CPSU and the Soviet state, of friendship and brotherhood of socialist countries',[10] while Party leader Josef Lenart announced that it had shown the advanced technology possessed by the socialist countries, their commitment to the peaceful exploration of space, the commitment of communist scientists to using technology for peaceful purposes, and the opposition of the Czechoslovak and Soviet people to the US 'neutron bomb' programme!

All the guest cosmonauts were carefully screened for political reliability, so that there would be no deviation from the chosen propaganda themes. Czech cosmonaut Remek, for example, was the son of the Czech deputy defence minister and had toured Czechoslovakia as a 19-year-old flight cadet in 1968 giving lectures on the virtue and legitimacy of the recent Soviet invasion of his country. Remek was followed by a series of guest cosmonauts from the other Warsaw Pact allies of Poland, East Germany, Bulgaria and Hungary. Significantly, cosmonauts from Vietnam, Cuba and Mongolia flew before the final Warsaw Pact country, Romania, saw its cosmonaut in orbit. This was a deliberate symbolic snub to Romania by the USSR, which had always been irritated by Romania's undeferential stance within the Warsaw Pact (it had refused to take part in the 1968 invasion of Czechoslovakia, for example), and the obvious gap before it was represented was a symbolic but firm slap on the wrist from Moscow.

After Hungary, the penultimate Warsaw Pact ally, it was Vietnam that flew the next joint mission with the USSR. Again, the flight was laced with political symbolism. The flight coincided with the Moscow Olympics, which the United States boycotted. Soyuz 37 saw Pham Tuan of Vietnam fly to Salyut in July 1980. Tuam was the first Asian to fly in space, and was a former air-force pilot who had shot down an American B-52 bomber during the Vietnam War. As 'mission specialist' he carried out an environmental survey of Vietnam from space, whose unsurprising conclusion was that Vietnam's environment had been very badly damaged by American military activities during the war, particularly by the use of the Agent Orange chemical defoliant. The propaganda message of the joint flight was very clear.

The next flight, Soyuz 38, took Arnaldo Tamayo of Cuba to Salyut. As a Cuban, and a military officer in the Cuban air-force, his flight was an important symbolic message of Soviet solidarity with the Cuban revolution and with the embattled Cuban socialist experiment. To rub salt in the wounds, Tamayo was also black and Soviet propaganda gently noted the irony that the Soviet Union had included a non-white in its space programme, although it had no black population, while the United States had never done so, even though more than 10 per cent of its population were of African descent. No accusation of racism was made, but the 'irony' was there for the whole world to see and draw the appropriate conclusions.

After Romania finally saw its cosmonaut in space, one from Afghanistan flew. Again, this was a highly political choice given the ongoing controversy, domestic as well as international, about the Soviet military presence in Afghanistan after 1979. But while the Vietnamese, Cuban and Afghan joint missions were perhaps unsurprising, the subsequent inclusion of cosmonauts from France and India was certainly a dramatic demonstration of Soviet inclusiveness regarding its space programme.

Political considerations remained clearly at the heart of both missions. Throughout the Cold War, notably during the *détente* period, it was always clear that political considerations determined both the nature and scope of space cooperation. It could be used to punish another state by reducing or refusing cooperation during a period of bad relations, or it could be used positively to symbolise an era of improving relations.

France, as a NATO member, was an unusual potential partner, but France was eager to demonstrate its autonomy and its freedom of national decision-making vis-a-vis the United States. The two countries already had a history of cooperation in space science going back as far as 1966, and the joint space flight was only one example of this cooperation.[11] With a national space programme of its own, and a leading role within the European Space Agency, France sought to avoid dependence on either superpower in space matters and felt that cooperation with the impressively successful Soviet programme would serve French national interests. The Deputy Head of the French space programme stated in 1981 that 'the Soviet Union is a great space power, which possesses immense technical and scientific possibilities… we are very satisfied with the development of this cooperation, if not for it we would have to substantially reduce our programme'.[12] Such comments were a validation of the Soviet programme and all the more dramatic given that 1981 was at the height of the 'Second Cold War' in Europe.

The relationship with India had to some extent grown out of the cooperation that both countries had with France, but the Soviet Union was in any case eager to develop its relationship with the most powerful state in the Indian Ocean region, particularly given the close relationship between Pakistan and China. Like France, India had its own robust national space programme, and was seeking to diversify its outside dependence, so that it too had much to gain from cooperating in space science research with the USSR. The Soviet Union in turn saw a cooperative relationship with India as something that would enhance its ties with one of the leading countries of the non-aligned movement, that would demonstrate the benefits of the Soviet model of socialism, and that might drive a wedge into the unity of the democracies. The joint flight in 1984 provided visible evidence of this successful cooperation.

The joint flight was politically significant for both countries. Agreement on the inclusion of an Indian cosmonaut in the Intercosmos programme had been reached in 1980, and the Indian Prime Minister met the Indian cosmonauts in training during an official visit to the Soviet Union in 1982. When Squadron-Leader Rakesh Sharma flew on Soyuz T-11 to the Salyut-7 space station in 1984 he brought with him 'an Indian flag, pictures of the country's political leaders, and a handful of soil from the Gandhi memorial'.[13] The following day he held a televised conversation from orbit, in English and Hindi, with the Prime Minister, Mrs Indira Gandhi.

In 1977, on the anniversary of Gagarin's flight, the Soviet army newspaper, *Krasnaya Zvesda*, declared that the USSR sought to achieve mutual benefits from international cooperation. 'Space research represents particularly broad opportunities in this respect. It is capable of uniting the efforts of different countries for the sake of solving the large tasks facing the whole of mankind to their mutual benefit. This has been graphically demonstrated by the successful implementation of the joint Soviet–American Soyuz–Apollo flight and by cooperation with scientists of France, Sweden, India and other countries'.[14]

For Oberg, the 1980s represented a 'decade of space exploitation rather than earlier space exploration'.[15] The USSR was seeking more practical gains from its space programme, in a sense initiating a new space race, not for prestige, or even curiosity, 'but for wealth and power'.[16]

Superpower cooperation

Karash calls the first half of the 1970s the most intriguing period in the history of US–Soviet space cooperation, since neither of the leaders of the two countries were initially enthusiastic about their national space programme, and neither initially saw space as a potential area for US–Soviet cooperation. This was particularly noticeable since both were openly looking to build such cooperation in other areas of the superpower relationship.[17]

The emergence of cooperation was also noteworthy because, as is discussed below, this period was one in which both countries were energetically pursuing the *military* uses of space in order to strengthen themselves in terms of their global strategic confrontation. In the second half of the 1960s, while the NASA budget was falling, the Pentagon's military space spending was increasing dramatically. Interestingly, at the height of the Cold War in the 1950s, two popular Soviet science fiction novels had depicted US–Soviet space cooperation. In both stories, an American astronaut had reached another world (the Moon in one, Mars in the other) and, on becoming trapped there, is rescued by a subsequent Soviet mission.[18]

As noted earlier, even during the 'space race' of the 1960s US Presidents Kennedy and Johnson had advocated the benefits of superpower space cooperation. In his 1963 speech to the United Nations, President Kennedy proposed a joint manned expedition to the Moon with the Soviet Union. A number of members of Congress subsequently protested to Kennedy about the proposal, arguing that it contradicted what the government had been consistently insisting, that the space race was a crucial part of the global confrontation with communism.[19] Kennedy defended himself by arguing that his administration had always been in favour of space cooperation with the Soviet Union and that such cooperation would only be possible if the United States had a strong space programme of its own. The technological, as well as the political difficulties, involved in such cooperation would have made it very difficult to implement quickly in practice, but Kennedy's arguments were consistent with his public utterances in the previous three years, and in the period following the Cuban Missile Crisis, the United States was seeking ways to increase cooperation between the two superpowers, as is evidenced by the signing of the Partial Nuclear Test Ban treaty in 1963. However, the Soviet Union did not respond to Kennedy's proposal, and in October 1963 Congress passed an amendment to NASA's annual funding bill which specifically forbade the United States government from agreeing to participate in a joint lunar mission with any 'communist, communist-controlled, or communist-dominated country'.[20]

President Johnson had also shown a willingness to encourage the idea of such cooperation, for example asking a 1965 press conference the question that 'as man draws nearer to the stars, why should he not also not draw nearer to his neighbor? As we push even deeper into the Universe, we must constantly learn to cooperate across the frontiers that really divide the Earth's surface'.[21] US administrations in the first two decades of the space age sought both to achieve American leadership in space, because this was felt to contribute significantly to US prestige and influence, and sought international cooperation, because it was believed that this would encourage other states to seek compromises with the United States on other issues in the hope of gaining access to American space technology.[22]

Space cooperation appeared because it ultimately came to be seen as serving the newly emerging interests of both countries. The early 1970s were characterised by a period of *détente* between the superpowers. A number of factors had come together to produce this 'era of negotiations' between the United States and the Soviet Union. The United States was struggling to cope with the political and financial pressures produced by the war in Vietnam, the need to implement social and economic reforms under the pressure of the racial and social tensions within the United States, the costs of alliance commitments in Europe and Asia, and the world-wide demands of the confrontation with the Soviet Union, particularly in the military–technology realm. An opening to improve relations with the USSR had also been made possible by the Ostpolitik foreign policy of West Germany, which had led to a significant improvement in West Germany's relations with the Soviet Union and its allies, and a consequent reduction in NATO–Warsaw Pact tensions. All these factors, plus fears of the dangers created by unconstrained arms racing, made the United States receptive to the idea of a limited rapprochement with the Soviet Union. President Nixon declared in his inauguration speech that 'after a period of confrontation, we are now entering an era of negotiations'.[23]

The USSR in turn was struggling to cope with the economic demands of its global commitments, faced an increasingly difficult relationship with its former ally China, which had seen heavy fighting on their eastern and western borders in 1968 and 1969, and was also aware of the dangers represented by the nuclear confrontation with the United States. A crucial additional factor was that by the beginning of the 1970s the Soviet Union had finally achieved nuclear 'parity' with the United States, a situation where the two superpowers had an approximate equilibrium in terms of numbers of strategic nuclear weapon systems, and relative capacity to inflict devastation on the other country. This equilibrium and its acceptance by the United States meant that the two sides were willing to codify their relationship as great power equals. Without this parity, and recognition by the United States, the Soviet Union would not have been willing to sign agreements that in effect locked it into a permanent position of inferiority relative to the USA.

Both countries sought to use this improved relationship to conclude agreements that reduced the dangers inherent in their Cold War relationship. The early 1970s were therefore marked by a series of arms control agreements, including the landmark SALT I Treaty of 1972. Some of these treaties, such as SALT, were hugely significant, others, such as the 1971 Seabed Treaty, were of no more than symbolic value. The symbolism was important, however, because the two countries were seeking a variety of ways both to dramatise their new relationship to the global audience, and to consolidate it by finding areas where mutually beneficial cooperation was possible.

One such area was space research. The symbolic importance of superpower cooperation in this area was seen as all the more significant and politically resonant, because of the drama of their intensely public competition and rivalry in this activity during the previous decade. For the Soviets, such cooperation also fitted in well to their revised foreign policy use of space in the post-1969 period. For them, it was an extension of the Intercosmos programme of international cooperation, which would see the beginning of joint missions with other countries three years later. The chairman of the Intercosmos Council argued that the 1972 US–Soviet space agreement was of immense political significance and showed that 'outer space is becoming, in all its aspects, an arena for broad international cooperation and demands joint efforts of many countries, especially those countries which have made considerable achievements in this matter'.[24]

In May 1972, The United States and Soviet Union signed the 'Agreement Concerning Cooperation in the Exploration and Use of Outer Space for Peaceful Purposes', agreeing to cooperate in a number of aspects of space science.[25] Political statements in both countries following the accords emphasised both the utilitarian benefit of space cooperation and its symbolic and practical importance as an alternative to the dangers of arms racing and war. Pravda declared that 'Earth is the planet of mankind. Cooperation in space paves the road to peace, mutual understanding and the good of all the people',[26] while US Senator Cook insisted that 'nations everywhere must begin to recognise that it is only through mutual interdependence that this world can exist peacefully for many tomorrows to come'.[27] TASS, the Soviet international news service, described the agreement as 'an important new act in the development of international relations'.[28]

The agreement made possible a number of important joint scientific programmes linking the two countries' space activities. From the American perspective the most useful was probably the three Soviet satellites launched in 1975, 1977 and 1978, which carried numerous American biological experiments. No comparable American unmanned biological missions or manned spaceflights were taking place in this period and the missions were therefore extremely valuable for US space biology scientists. During the same period the two countries coordinated their Venera and Pioneer-Venus missions to Venus and exchanged the resulting data.[298] Some leading officials within NASA even speculated during this period about the possibility of developing a space station jointly with the Europeans and the Soviet Union.[30]

The most dramatic symbol of the new desire to cooperate was the Apollo–Soyuz Test Project, under which the two countries committed themselves to a joint manned mission in which their spacecraft would link-up and dock in space. The idea of the 'hand-shake in space' had come from Secretary of State Henry Kissinger, who saw it as an effective way to symbolise the *détente* foreign policy being pursued by the Nixon administration.[31] The ASTP obliged the two states to cooperate closely in the three years leading up to the mission, in order to allow each to understand how the other's systems operated, and to train the astronauts and cosmonauts for the joint mission. The space link-up took place on 17 July 1975.

One of the crucial factors that made such a joint mission possible was that by the beginning of the 1970s the Soviet Union had gradually begun to reduce some of the secrecy surrounding its space programme. The military dimension remained hidden, but secrecy was less prominent for the manned and unmanned scientific missions. This was a significant development. As Oberg noted, 'the most salient feature of the old Soviet space program was itself invisible. It was the secrecy that Moscow wrapped around all its activities – a secrecy designed to mislead, either to allow protection of real technology or to trick foreigners into overestimating the level of Soviet space technology'.[32] The greater openness was a reflection of the Soviet Union's growing confidence in the genuine achievements of its space programme. But it came at a price; when disasters such as the loss of the Soyuz 11 crew in 1971 occurred, they could not be hidden.

The Apollo–Soyuz joint mission was the last manned American spaceflight for six years. Americans would not fly in space again until the launch of the first space shuttle in 1981. The United States expected the Soviet Union to capitalise on this period of relative American inactivity. The US believed that periods in which the USSR appeared to dominate space exploration would be characterised by efforts to maximise its unilateral propaganda gains. A Senate report in 1982 declared that 'a surging Soviet space program pitted against an American programme in retrenchment, particularly in an environment of deteriorating political relations, could create irresistible political opportunities for the Soviets to play space politics with even greater intensity'.[33]

By the second half of the 1970s the *détente* relationship was beginning to cool once more, and while the advantages of cooperation meant that it continued in some areas of space research, once again political realities shaped the limitations of that cooperation. What this period demonstrated was that space cooperation, like arms control, was a reflection of improving interstate relations, not a generator of them. If relations were improving, cooperation in space could be used to symbolise dramatically and encourage that improvement, but when relations were poor, space cooperation was simply not politically feasible. At all times, therefore, such cooperation remained hostage to broader political dynamics. The end of the *détente* period, and a re-emergence of the Cold War confrontation in the early 1980s made this dramatically clear.

President Reagan's 1983 Strategic Defense Initiative was in this, as in many other respects, a critical development for the Soviet Union. SDI was dependent on the development of highly advanced technology, much of which would be deployed in space. The American willingness to pursue such a programme in the face of Soviet objections demonstrated not only that the *détente* period was clearly over, but also that a very different conceptualisation of the possibilities of US activities in space was dominant in Washington. The following year, Mikhail Gorbachev became Soviet leader and sought to bring about a distinct change in the foreign policy orientation of the USSR. Gorbachev emphasised a Soviet desire for greater cooperation with the United States, and sought to reduce the obsessive focus on the superpower relationship at the expense of other important areas of the world.[34] Nevertheless, the desire to use symbolically important propaganda was evident in Gorbachev's decision to rename the latest in the Salyut space station series, so that instead of being Salyut 7, the station, launched in February 1986, was named *Mir*. The use of a Russian word for 'peace' was designed to suggest the peaceful nature of the Soviet programme and to draw attention to the American effort to militarise and 'weaponise' space through the Strategic Defense Initiative. The SDI needed to be challenged symbolically in this way because Gorbachev was aware that a Soviet effort simply to match the American programme would not only be strategically destabilising, but was likely to expose the economic weaknesses and technological limitations of the USSR. If 'the concept of an unending, mortal struggle for survival in a hostile environment is a key to understanding Soviet foreign policy',[35] then the challenge to compete in building a space-based ballistic missile defence system was the 'bridge too far' which confirmed Gorbachev in his conviction of the need for a fundamental reorientation of Soviet foreign policy.

US policy was also undergoing a revision during the same period. The first Reagan administration had broken off virtually all space cooperation with the USSR following the Soviet move into Afghanistan, the imposition of martial law in Poland, and the re-emergence of a Cold War relationship between the two superpowers. In 1982, the 1972 US–Soviet agreement on space cooperation expired. The original agreement was renewable at five year intervals, and President Carter had renewed it in 1977. However, the Reagan administration allowed the agreement to come to an end, demonstrating once again the absolute linkage between superpower space cooperation and prevailing diplomatic and political relations during the Cold War period.

This was a dramatic move, since by the early 1980s a significant amount of cooperation between the USA and USSR in space exploration had become almost routine. A positive approach towards international cooperation in general had been part of the remit of the American space programme since its inception. While considerations of power, prestige and military effectiveness were the drivers of the US programme, the National Aeronautics and Space Act of 1958 had committed NASA to the pursuit of cooperation with other countries, identifying such cooperation as a fundamental goal of the American space programme. In his first State of the Union message, President John F. Kennedy declared that his administration intended to 'explore promptly all possible areas of cooperation with the Soviet Union and other nations in space matters'.[36]

International space cooperation was seen as a useful tool for pursuing a range of American foreign policy goals, including the reduction of international tensions, particularly through greater transparency and access to information, increasing US prestige through the high profile of the American technological and organisational achievements in the space programme, promoting economic development and making possible increased political access to countries where American influence was otherwise limited.[37] In addition, cooperation was seen as a way of using space to increase American prestige. NASA administrator James E. Webb argued that international cooperation projected an image of a United States 'wanting to work with other nations to develop science and technology, the image of a nation leading in this field and willing to share this knowledge with other nations'.[38] His predecessor as administrator, T. Keith Glennan, had felt that international space cooperation might even prove of fundamental importance in producing a permanent thaw in the Cold War, speculating that from such cooperation 'may yet come that common understanding and mutual trust that will break the lock step of suspicion and distrust that divides the world into separate camps today'.[39]

Other observers of NASA efforts at international cooperation were less sanguine, however. Cash argued that throughout the Cold War period, NASA was never interested in genuine cooperation with the Soviet Union, since 'the primary goal of NASA has never been remotely related to cooperation; it has been to beat the Soviets into space...the very notion of using space cooperation to create a new political reality would be inconsistent with this conception'.[40]

During its passage through Congress the Space Act was amended to increase the attention paid to the utility of international cooperation on the grounds that 'international space cooperation could promote peaceful relations among states and form the basis for avoiding harmful and destructive actions in space'.[41] The Act's preamble declared that 'the Administration, under the foreign policy guidance of the President, may engage in a program of international cooperation in work done pursuant to this Act'.[42] An amendment to the Act passed in 1975 further directed NASA, in consultation with the President and Secretary of State, to 'make every effort to enlist the support and cooperation of appropriate scientists and engineers of other countries and international organizations'.[43] On the basis of such authorisation, NASA has historically cooperated with other countries and international organisations in a range of fields, including providing satellite launches, provision and analysis of space-derived data, and joint experiments, including provision for foreign payloads on US scientific satellites. NASA has also placed payloads on foreign spacecraft such as missions by the European Space Agency. Such cooperation was not limited to American allies in the developed world. NASA cooperated with the Indian Space Research Organisation in the 1975 Satellite Instructional Television Experiment (SITE), in which the NASA ATS-6 communications satellite was used.

Cooperation with the Soviet Union was also a feature of the American space programme from 1962 onwards, despite this being the most intensive period of the superpower manned space race. When Khruschev congratulated Kennedy on the successful flight of John Glenn, the first American to orbit the Earth, he suggested that if the two countries pooled their efforts in space exploration and use, 'this would be very beneficial for the advance of science and would be joyfully acclaimed by all peoples who would like to see scientific achievements benefit man and not be used for Cold War purposes and the arms race'.[44] Kennedy responded positively, though the two countries agreed to limited coordinated activity, rather than any merging of their programmes along the lines that Khruschev seemed to have been suggesting. Initial cooperation focussed on meteorological studies and was subsequently expanded to include telecommunications experiments and geomagnetic mapping. The most dramatic example of US–Soviet cooperation during the First Cold War was the 1975 Apollo–Soyuz joint flight. It had been intended that the joint Apollo–Soyuz docking unit developed for this mission would make future joint missions possible. However, the ASTP turned out to be the final flight of an Apollo spacecraft, and its successor, the Space Shuttle, did not have such a docking mechanism, and none would be developed prior to the end of the Cold War. Apollo–Soyuz in the event was an episode, rather than a harbinger of future cooperation.

So strong was the idea of space cooperation as a bridge between the wary superpowers that as early as 1984, when President Reagan began tentative efforts to improve the dangerously poor relationship with the Soviet Union, space was immediately identified as an area where a dramatically symbolic move might be made. By the spring of 1984, the Senate Foreign Relations Committee was recommending a renewal of US–Soviet space cooperation, and in July, President Reagan proposed a joint US–Soviet manned mission which, like the 1975 mission, would test the feasibility of each country being able to rescue the stranded astronauts of the other. A bill to restore superpower space cooperation was signed into law by the President in October 1984.

NASA was not the only US agency promoting the use of space in international cooperation. The US Agency for International Development (AID), was also exploiting the technology to promote development in the Third World, particularly for delivery of services to remote populations. The success of the joint US–Indian SITE experiment led AID to fund a follow-on project involving 27 countries.[45] The State Department was involved in the coordination of the diplomacy involved in the provision of Earth remote-sensing LANDSAT data to foreign remote sensing programmes. LANDSAT was perceived by the United States as much as a diplomatic as a technological tool in which 'the United States has also used it as an ambassador for US space technology by selling data on a public non-discriminatory basis, and through arrangements for direct transmission of LANDSAT data (on a fee basis) to foreign-owned and foreign-operated ground stations'.[46] AID was also involved in the provision of US aid in the delivery of remote sensing programmes in developing countries. International training for such programmes was also provided by the US Department of the Interior.

The United Nations has also been a key player in the encouragement of international space cooperation and has played a key role in the regulation of space activity since the dawn of the Space Age, establishing the UN Committee on the Peaceful Uses of Outer Space (COPUOS) in 1961. COPUOS has been central to the formulation of international treaties which form the basis of the international law of outer space. These have included agreements on the rescue and return of astronauts (1968), responsibility for damage caused by space objects (1972), registration of objects launched into outer space (1974) and activities carried out on the Moon and other celestial bodies (1979). The most important agreement, however, is the 1967 Outer Space Treaty which established the crucial understanding that outer space and celestial bodies are not subject to national appropriation and sovereignty, and that states are held responsible for their own space activities and those of their citizens.[47]

This crucial clause meant that space became a domain that could not be dealt with in terms of the categorisations familiar in terrestrial international relations. The contemporary international system in the second half of the twentieth century was one in which the fundamental basis for interstate relations were the concepts of sovereignty and territoriality. States were sovereign within their own territorial jurisdictions, and by the 1960s those jurisdictions covered the entire land surface of the planet. International law at this time was based on the assumption that states exercised clear and undisputed control over fixed territorial areas covered by their borders and other states had no legal right to intervene within the territorial boundaries of the sovereign state.

At the outset of the space age there were real concerns that the principles of sovereignty and territoriality would simply be extended beyond the Earth's atmosphere, just as they had been extended throughout the world from the seventeenth century onwards. This would have been understandable in some ways given that there were no real precedents for the new situation being created by the exploration of space, though there were some limited parallels with the legal regime that covered the oceans beyond national state jurisdiction. Where such parallels were instructive, for example, was in encouraging the idea that resources existing beyond the Earth's atmosphere formed part of the 'common heritage of mankind'.

The belief that space would in time be covered by legal principles very similar to those operating on Earth in turn encouraged the idea that Earth-orbital space, and planetary bodies beyond it, would become appropriate sites for the establishment of military bases, where weapons of mass destruction could be targeted against terrestrial states.[48] In reality, the Moon would have been a ludicrously inappropriate and counterproductive place to base nuclear weapon launchers. The same was not necessarily true, however, of near-Earth space, where there was considerable early interest in the exploitation of the zone for military purposes. USAF General Curtis Le May argued that 'the present area of military interest is within the sphere bounded by the synchronous orbit'.[49]

In 1961 the General Assembly of the United Nations unanimously adopted a resolution, in which it laid down two key principles. First, that international law, including the United Nations Charter, applied in outer space. On its own this declaration would have left space open for the extension of existing international legal practices such as sovereignty and territoriality. However, the same resolution embraced the key assertion that 'outer space and celestial bodies are free for exploration and use by all states in conformity with international law, and are not subject to national appropriation'.[50]

When the novel environment of space interacts with terrestrial polities in a traditional manner it can seem amusing. In 1995 a joint Russian–American crew were sent up to the Mir space station on a Russian Soyuz spacecraft.

Four months later they returned to earth on the space shuttle *Atlantis*. When US Customs and Immigration service discovered that this was going to happen, they insisted that the two Russians would need visas to enter the United States from Earth orbit, and the Atlantis crew had to take the visas up to them![51]

International cooperation is also reflected in the range of functional cooperative activities coordinated by the various Intergovernmental Organisations that have regulatory responsibilities in the space domain, such as the International Telecommunications Satellite Organization, the International Telecommunications Union, the International Maritime Satellite Organization, and the World Meteorological Organization. The WMO, for example, has presided over a system of voluntary free interchange of meteorological data, both terrestrial and satellite derived, between the member countries. These functional operations lend themselves to analysis through liberal international relations theory. They make possible the effective utilisation of the communications frequencies used by satellites, the allocation of positions in the geostationary earth orbit, and so on. Technical cooperation of this kind reflects the fact that space cooperation is produced not only by a desire to exploit political symbolism, but also by the technical necessities of space operations.[52] Cooperation of this kind has allowed a more rational exploitation of Earth orbit and exploration of the Solar System, though the overlap between the programmes of different national agencies, which are driven by the differing political needs of their government sponsors, is still clearly visible. Even after the successful joint NASA–ESA mission to Titan the space scientists involved asked 'why two missions to Mercury, one from NASA, one from ESA? Why two agencies with missions to Mars? Why can't we explore the Solar System together?'[52] The answer, as always, is politics. Space programmes fulfil political as well as scientific functions, and these are often best served by national missions that focus attention and prestige on the country that initiates them.

Chapter 5

European integration and space

One programme where purely national considerations have been significantly sublimated into a joint international effort is in the work of the European Space Agency (ESA) and its predecessors. Although a European space programme did not get underway until the 1960s, Europeans played a prominent part in the development of space science in the first half of the twentieth century, with notable contributions from scientists and engineers such as the Russian Konstantine Tsiolkovsky and the German Werner von Braun. Indeed, both the American and Soviet programmes benefited from the capture of German V-2 experts at the end of the Second World War in Europe in 1945. The eastern half of Europe, through the USSR, dominated the initial phase of space exploration.

Western Europe, in contrast, was slow to develop a space programme, though several individual nations had programmes involving various aspects of space research, particularly the United Kingdom and France. Nevertheless, a number of forces helped push the Europeans towards collaboration in this field at this time. One factor was the general dynamic favouring European cooperation during the 1950s. The Treaty of Rome, which created the European Economic Community, was signed on 25 March 1957 and marked a major acceleration in the momentum for European cooperation, collaboration and integration in the economic sphere. The space age was initiated less than six months later with the launch of Sputnik I.

European space cooperation was self-consciously pursued as a form of functional cooperation, designed to bring political as well as scientific and technological benefits. The theory of functionalism, which has underpinned the European integration project, 'is based upon the hypothesis that national loyalties can be diffused and redirected into a framework for international cooperation in place of national competition and war'.[1] For theorists of functional integration, such as David Mitrany, the objective of the integration process was to emphasise international cooperation in the social and economic fields and to create gradually a range of supranational institutions which would take over these tasks from national governments.[2] Mitrany believed that the functionalist approach could bring nations together,[3] and that scientific and technical cooperation should be prioritised because it was less politically contentious than other areas.[4]

The relationship between this general process of European functional integration, and the specific construction of the European space programme was highlighted in 1989, at a ceremony celebrating 25 years of European space cooperation. ESA Director-General Reimar Lust concluded his address by referring to the inspiration he had always derived from the words of Robert Schuman, one of the original architects of European integration, that 'Europe will not be achieved overnight, and not all at once. It will gradually come to exist as a result of practical achievements which in the first instance give rise to real solidarity'.[5]

The idea of exploiting the potential of space was a profoundly important one. Neo-functionalist theorists such as Haas identified the main impetus behind integration as being economic self-interest, so in any successful integration project the participants needed to perceive the likelihood of tangible benefits from the process of collaboration.[6] Many of the problems experienced by the European space effort in its first decade came from the difficulty of finding structures and procedures that would clearly meet the requirements of economic self-interest.

Although functionalism has clearly been a factor in the development of European space cooperation, encouraging an analytical model shaped by liberalism, in fact much, if not most, analysis of this policy area in the international relations literature has been from a neo-realist perspective, emphasising rational state behaviour, the protection of national autonomy and the central importance of intergovernmental bargaining.[7]

However, the neo-realist approach is of limited value in explaining the form and process of European space cooperation. In particular, it pays insufficient attention to the ideological aspects of the cooperation, to the concept of European as distinct from national autonomy, to the crucial importance of non-state actors in this area and to the significance of institutions as 'an independent variable, influencing the process of preference formation'.[8]

Origins of ESRO and ELDO

The origins of the joint European space effort were very much influenced by initiatives by leading scientists such as Massie in the UK, Auger in France and Amaldi in Italy. These initiatives were important because, amongst other things, they helped to shape the European avoidance of collaborative military space which became characteristic of subsequent integration. The scientists were members of an important epistemic community, who were able to influence national interpretations of state interests, and increase the likelihood of convergence in state behaviour at the international level. West European intergovernmental cooperation in the scientific field had already begun with CERN, the European organisation for nuclear research established in the mid-1950s, and Euratom, which was being developed as part of the efforts to create a European Economic Community.[9]

France had developed the small *Veronique* sounding rocket at this time, but only Britain had followed the superpower lead in developing long-range ballistic missiles, with the *Blue Streak* and *Black Knight* programmes. However, *Blue Streak* was cancelled as a military launcher in April 1960. Only two months later the European Group for Space Research was founded. The coincidence of timing was important since the *Blue Streak* cancellation left the British government wondering if there were any way it could justify the large sums that had already been spent developing the missile, and the possibility of converting it to a satellite launcher was soon raised.

At this time there were important moves within NATO to shape the future development of European space collaboration. In June 1957 NATO set up a Task Force on Scientific and Technical Cooperation, and a Science Committee, which recommended a NATO space programme. The NATO Heads of Government conference in December 1957 established a Consultative Group on Space Research.[10] The NATO Secretary-General's science adviser meanwhile recommended the creation of a European NASA to work in partnership with its American equivalent.[11] It was against this background of NATO interest in European space activities that European scientists sought to promote space cooperation for purely 'peaceful' purposes.

In April 1959, Amaldi noted that space research was being monopolised by the two superpowers and called for the creation of an international organisation of European countries to enable Europe to participate. The proposed European Space Research Organisation he argued, 'should have no other purpose than research and should therefore be independent of any kind of military organisation and free from any official secrets act'.[12] This would enable it to maintain what he referred to as its 'moral authority' and also enable a wide cross-section of European states, including the neutral states outside NATO such as Sweden and Switzerland, to take part. Amaldi was insistent that the European space organisation should be civilian in character, and not linked with any military organisation, a demanding feature given that all space exploration up to that point had been driven by military rationales and carried out with military involvement.[13]

Amaldi believed that the European organisation should have its own launch site and develop its own launch vehicle, since 'if the military maintained a monopoly on the construction of rockets, each country would build its own'. Despite his concerns about the military domination of space, Amaldi and his colleagues were aware that with space, the boundaries between research and development, including commercial developments, and between peaceful and military uses, were not hard and fast and were difficult to keep separate. This was another theme that would become central to the development of European space activities.

The desire of the West European states to be active in space exploration seemed at first to be more due to a fear of being left behind in a new activity the superpowers clearly deemed important, rather than a result of a clear perception of the future importance of the activity *per se*. The European governments at this time believed that several possible options were open to them in regard to space. One was simply to ignore it and do nothing. Even as early as 1960 this seemed a dubious option, because the long-term potential of space in various realms was beginning to be realised. A second possibility was for the larger and more technically advanced states to pursue individual programmes. This was a tempting option for some states, in terms of the benefits for national prestige and for military spin-offs which would contribute to national independence. Against this, however, were the enormous costs that would be associated with such a national programme.[14] For the smaller states, such an avenue was unrealistic, but both Britain and France were in a position to give it serious consideration. However, they were aware that their programmes would seem modest compared to those of the superpowers and worried about the proportion of scientific talent that might have to be tied up in such a programme, to the detriment of other projects, such as the development of nuclear capabilities. Nevertheless, it was significant that in a key preparatory meeting of European space scientists held in October 1960, it was emphasised that the proposed new European space organisation would not replace or compete with national programmes, but would complement them and 'enhance their efficiency'.[15]

A third possibility would be to submerge all national capabilities in a single United Nations space programme. While this would have advantages in terms of sharing costs and expertise, the realities of international politics at this time seemed to rule out this option in practical terms. Given the superpower antagonism, reflected in the deadlock in the UN, and their inability to trust each other or cooperate effectively, a UN programme would be impossible to organise, even if the superpowers were prepared to share the propaganda advantages of a joint programme, which seemed highly unlikely.

A more practical option would be for Europe to cooperate specifically with one of the superpowers in space research, which, given the geopolitical realities of the time, meant the United States. In March 1959 the US had announced its willingness to launch scientific satellites on behalf of other nations. This was a tempting offer for the European scientific community, who were interested in getting their experiments into space, rather than worrying about the political and strategic implications of European collaboration.[16] But while this option held attractions, there were many factors arguing against its adoption. Europe did not have an absolute requirement for such cooperation, because it already possessed a potentially effective satellite launcher of its own, the British-developed *Blue Streak* missile. Moreover, many of the economic and technological benefits derived from the challenge of developing a space launcher would be lost if Europe simply chose to use

American rockets. Nor were all European governments sure that the United States would always be willing to make such services available in the future. There was a risk that at key moments the United States might decide that it was not in its own national interest to make launchers available to their commercial or political rivals in West Europe. NASA's founding Act had stressed international cooperation for peaceful applications as a fundamental aspect of American space policy. Nevertheless, subsequent US policy on space cooperation with Europe saw one of its key purposes as being the demonstration of American political leadership of its allies by providing launch facilities that alone made their cooperative satellite ventures feasible.[17] Finally, if cooperation with one or more superpower could bring benefits, then there was no reason why the existence of an independent European programme should necessarily exclude the possibility of such cooperation.

These considerations pointed Europe towards a preference for a programme where the West European states cooperated between themselves as a group. Numerous advantages could be identified as flowing from this strategy. In the first place, the enormous costs of space exploration and utilisation would be shared between a large number of countries, making them affordable. For the same reason, the pressure on any one country's limited pool of scientific and engineering talent would be reduced. At the same time, bringing European specialists from many different countries together might generate a 'cross-fertilisation' of ideas and produce greater progress than any single country acting on its own could achieve. For West Germany, there was the additional benefit that it would be a vehicle for re-entering the space research field. Germany had been forbidden long-range rockets after 1945 because of its use of the V-2 missiles during the Second World War. German industry and academia believed that 'the lost political legitimacy of German space activities could be restored only by accepting the roundabout 'European' road'.[18]

Politically, it would help to unite Western Europe, since it would bridge the existing divisions between the NATO and neutral states, and between the EEC and EFTA countries. Britain's Minister of Aviation stressed to his European counterparts the 'immense political advantages in Europe getting together on a project of this kind which would straddle the existing divisions between Six and Seven'.[19] And if successful, a joint European space programme would increase Europe's power and her influence in the world. The Consultative Assembly of the Council of Europe was one of the bodies calling for the creation of a European space programme in order to enable the industries of the member countries to take part in 'the commercial developments resulting from space technology; and to ensure the application of this new knowledge to the benefit of their peoples and their economies'.[20] To that end, the Assembly recommended that member states should 'study as a matter of urgency the possibilities and cost of setting up a European agency to undertake a space programme, based upon a space vehicle developed and built in Europe, and to promote the peaceful uses of outer space'.[21] These recommendations reflected those put forward in a crucial document that Amaldi had circulated to key European figures early in 1959. The document, *Space Research in Europe,* emphasised the non-military character of the proposed organisation, which should be modelled on CERN.[22]

ELDO and ESRO

At the time that European governments were investigating the possibility of cooperation in space, the United Kingdom was in the process of abandoning the military *Blue Streak* missile programme. However, although it had decided that for various reasons the missile was no longer an appropriate choice as a long-term deliverer of British nuclear weapons, the government was anxious to avoid writing off all the development costs of the programme, and was therefore keen to offer the missile as the basis for a European satellite launcher. France meanwhile was completing development of the *Veronique* rocket, capable of launching small satellites. The technological basis therefore existed for the development of a heavy satellite launcher using the British and French missiles as the first two stages. Significantly, the Europeans decided not to pursue the possibilities of the new organisation being sponsored by the UN, or even by the Organization for European Economic Cooperation, because this would allow non-European participation (the USA and Canada were about to join the OEEC), and reduce its impact as a vehicle for European cooperation. Also important was the fact that Switzerland took the lead in hosting the key international meeting of government representatives that created the working groups who drafted the founding documents for the proposed organisation. From the outset, therefore, the leading role played by the neutral states helped shape the future direction of European space cooperation. At that same meeting, Britain and France proposed that the development of launchers should be the responsibility of a separate organisation, a proposal that the scientific community supported.[23]

At a conference held in February 1961, Britain and France committed themselves to jointly developing a three-stage launcher capable of orbiting a one-tonne satellite. A follow-up international conference in November 1961 led to the drafting of the Convention of the European Launcher Development Organisation, which was signed in March 1962 by France, The United Kingdom, Italy, Belgium, West Germany, the Netherlands and Australia. The somewhat surprising presence of Australia in an overtly European organisation was made necessary by the fact that the only available launch site was the Anglo–Australian testing and launch site at Woomera in Australia. Because of the advantages of launching rockets from bases on the equator, in 1966 ELDO decided to begin construction of an equatorial launch site at Kourou in French Guiana.

From the outset ELDO was hampered by the fact that the different participating governments were driven by significantly different policy logics, economic for Britain, political for France and technological for Germany.[24] Britain's promotion of *Blue Streak* as a European satellite launcher was in part driven by a reassessment of attitudes towards European integration that had taken place in 1960. As the European Economic Community began to appear a successful initiative after 1957, the British government came to regret its initial decision not to participate in the organisation. Offering leadership in the field of European space cooperation was 'one dimension of a wider strategy aimed at closer integration with the Six, and was seen in Whitehall as an important proof of Britain's (new) European credentials'.[25] This perspective was echoed by West Germany, whose government welcomed the British proposal for a European launcher organisation as 'an opportunity to strengthen the linkages of the United Kingdom with the continent as such, and also as a possible first step towards an enlargement of the EEC'.[26] In a conversation between British Prime Minister Macmillan and German Chancellor Adenauer in February 1961, Macmillan emphasised the political advantages of the proposed joint European satellite launcher and Adenauer declared himself totally in agreement with Macmillan's view.[27]

The ELDO Convention committed the signatories to the development and operation of a space launcher and ancillary equipment. It stipulated that ELDO should concern itself only with peaceful applications of the launchers and equipment. Britain had already assured the US government that when *Blue Streak* was converted to a civilian launcher, all its specifically military features would be removed. This was important because the US did not want to encourage West Germany or France to acquire an IRBM capability.[28] The civilian version would not have an inertial guidance system or re-entry properties and would not draw on classified American information.[29]

It was felt that if Europe was to seek genuine autonomy in space, then it clearly needed its own launch rocket. The fact that Europe had to rely on NASA to launch her satellites was seen as a major weakness. Britain would provide the first stage of the joint European launcher using *Blue Streak*, France would provide the *Coralie* rocket as the second stage and West Germany was given responsibility for developing the third stage (*Astris*) of the *Europa* rocket. Italy was primarily responsible for developing the satellite test vehicles. The organisation's headquarters were established in Paris.

ELDO was beset with major difficulties throughout its history. The three-stage multinational rocket programme was not a success. While the French and British rockets were efficient systems in themselves, it proved difficult to mate the two systems on a single rocket. The German third stage was not a success. Germany had no existing missile development programme. While she had pioneered the development of such technology, those pioneers were now working for the American and Soviet space programmes. Germany, unlike Britain and France, was assuming a major commitment without the resources or experience of a pre-existing national space programme to draw on. Experience demonstrated that a robust national programme was 'a precondition for an effective cooperation on the European level'.[30]

ELDO also experienced severe budgetary problems. Lacking experience in such programmes, Europe significantly underestimated the development costs, and the initial budget allocations were quickly exposed as being quite inadequate. But this, in tandem with the problems encountered in developing the technology, led many of the member states to reassess their commitment to the programme. By 1965 the organisation was in crisis and an emergency conference had to be held. Britain, which played perhaps the leading role in ELDO and contributed the largest percentage of the ELDO budget, initially argued that the organisation should press ahead with more ambitious goals if the financial situation was to be turned round. However this period of European history was marked by major disputes between Britain and France. Britain resented the continuing veto by France of its application to join the European Economic Community, while France was bitter about the imposition of the Flexible Response doctrine on the NATO allies by the United States and saw Britain as an untrustworthy Anglo-Saxon trans-Atlantic state with no real commitment to the cause of European unity. France therefore responded negatively to the British ELDO proposals and countered by arguing that ELDO activity should be reduced rather than accelerated.

ELDO's difficulties were compounded by the policy of *'juste retour'* enshrined in the Convention. Under this philosophy, profits and jobs created through the programme were to be allotted to the member states in proportion to the financial contribution they had made. However, with the *Europa* programme this was difficult to achieve, since the main elements had been allocated to a limited number of countries, and for reasons of efficiency it often made more sense to distribute contracts through competitive tendering rather than allocate them to a country that did not actually have the technological capability to deliver a cost-effective product. Disputes over the fairness of the way in which *juste retour* was being implemented damaged the working relationships between the member countries, but were extremely difficult to overcome within the limitations of the programme.[31] The ELDO Convention contained a major flaw in that it provided no institutional mechanism for political discussion within the organisation.[32] Nor was there an effective mechanism for coordinating the activities of ELDO and ESRO. The first conference of ELDO ministers, held in July 1966, emphasised the need for close and effective coordination between the European space organisations, and called for the creation of a European Space Conference to gradually harmonise European space activities.

The first such ESC took place in December 1966 and was attended by all the ELDO states, plus Denmark and Spain, with Austria, Greece,

Ireland, Sweden and Switzerland attending as observers. The second ESC took place in July 1967. It was attended by all the ELDO and ESRO states. The Conference agreed that the ESC would become a permanent body and would meet once a year at ministerial level to work out and ensure implementation of a coordinated European space policy. Although the ESC was a fairly informal body, it nevertheless had some significant successes in finding solutions to some of the budgetary and programmatic disputes that were hampering progress in ELDO and ESRO.[33]

The continuing development problems of the Europa rocket, combined with Britain's worsening economic situation meant that by 1966 Britain too had become sceptical of ELDO's value. The funds allocated to Europa proved inadequate. In 1966 ELDO committed itself to developing a more powerful Europa II, capable of lifting 200 kg to geo-stationary Earth orbit. By 1966 however, Britain, which had been the strongest advocate of Europa, had begun to have serious doubts. A Foreign Office document issued in February 1966 expressed doubts about the technological use and economic viability of the programme. It claimed that ELDO was developing an obsolescent and uncompetitive launcher which would not bring economic benefits to justify the money spent developing it, and suggested that the Europa II was unlikely to change this situation.

The critical interrelationship between European space policy and the wider politics of European integration was demonstrated in 1967. For several years prior to this, Britain, seeking to join the European Economic Community had been trying to play the part of the good European. However, in November 1967, President de Gaulle of France delivered another devastating veto on Britain's application for membership. The British government now saw no further point in pursuing a generous approach towards France and almost at once began to oppose French policy vigorously in several key areas. One of them was European space policy, where the UK blocked French-led efforts to reform ELDO and significantly increase its budget.[34]

In 1968 the UK government announced that it would not continue to fund ELDO after 1971 and justified the move by declaring that all UK research programmes dependent on government support must meet the criterion of 'economic justifiability'. Continuing economic problems left ELDO in a state of permanent crisis thereafter. The troubled Europa rocket was abandoned in favour of a new launcher, Ariane, which would be developed by a single lead-state, France. This left ELDO effectively redundant.

A fundamental political problem for ELDO was that it was caught up in the internal political tensions which bedevilled the other major European economic and security organisations in the mid-1960s. There was a profound political opposition between France and the United Kingdom on basic international political strategy. France wanted to achieve genuine autonomy in space for Europe, and saw avoidance of dependence on NASA and the United States as a crucial objective for a European collaborative programme. Britain, in contrast, simply wanted to utilise existing British technology to maximum effect, and for Europe to 'pull its weight' in the North Atlantic Alliance. It was opposed to any initiative that might weaken the links between the United States and its European allies. West Germany saw the issue as a pragmatic one, and was happy to use NASA launchers, if that was the most cost-effective option. Indeed, it hinted that it might withdraw from ELDO if it could obtain NASA guarantees of launch facilities for national payloads.[35]

Given these fundamental differences in political outlook between the countries building the three component parts of the European launcher, it was hardly surprising that the project failed. By 1970 ELDO had not managed to launch a single satellite with the Europa rocket, nor had it achieved anything which would have been beyond the financial and technical capabilities of the larger member states. One knowledgeable American observer of the European programme suggested that 'the conception, development and demise of the ELDO must contain a series of lessons, often painful, on how not to organise, fund, manage and operate a cooperative technical effort which crosses international boundaries'.[36] ELDO's activities ceased in May 1973 and its assets and staff were transferred to ESRO.

At the same time that ELDO was being established, a second key European organisation came into existence. In December 1959, at the initiative of the French government, a protocol was signed establishing a Preparatory Commission on Space Research, which was tasked with creating the framework for a new European Space Research Organisation (ESRO). The ESRO Convention entered into force in March 1963. Under the terms of its convention, ESRO was to promote collaboration among European states, exclusively for peaceful purposes. The initial member states were Belgium, Denmark, France, West Germany, Italy, Netherlands, Spain, Sweden, Switzerland and the UK. Austria, Norway and Ireland were given observer status.

The first four years of ESRO saw the successful establishment of a number of facilities required for the European space programme to operate effectively. These included the ESRANGE sounding rocket launch-site at Kiruna in Sweden, the ESDAC/ESOC data and documentation centre at Darmstadt in West Germany, the ESRIN space physics laboratories at Frascati in Italy and the ESTEC technology centre at Noordwijk in the Netherlands.

The initial objectives of ESRO focussed on launching small sounding rockets to the upper atmosphere and small satellites to near-Earth orbit. The budget was set at $306 million to cover the medium-term programme of eight years' activity. The budget would be covered by annual contributions from member states assessed on the basis of national income. While ESRO was successful in establishing its infrastructure, the financial resources committed in 1963 to cover an eight-year programme were rapidly shown to be hopelessly inadequate. Owing to inexperience in such activities,

ESRO, as ELDO had done, vastly underestimated the costs involved in the programmes. As a result some missions had to be abandoned.

Nevertheless, unlike ELDO, ESRO can be seen to have been relatively successful. It was set up to encourage European collaboration in space research and technology, and to a large extent it succeeded, in that it managed to achieve a significant degree of autonomy in its organisation and in the implementation of its programmes. The 1968 (NASA-aided) launchings of the ESRO-I, ESRO II and HEOS I satellites heralded the arrival of Europe as a space power.[37]

A more fundamental political crisis arose in 1970 when certain member states strongly urged that ESRO's operations be extended from work on purely scientific satellites to include the development and launching of applications satellites, that is satellites such as communications satellites launched for commercial rather than scientific purposes. This proposal had been hovering in the background for several years. In November 1972 the ESRO Convention was amended to make this possible, and the amended ESRO Convention formed the basis for ESA's Convention after the December 1972 decision in principle to set up the successor organisation.

From 1968 to 1970 the ESC meetings had failed to produce agreement on a coordinated European space programme. This was because of fundamental disagreements over the European launcher programme, the desirability of cooperating to produce application satellites, and the extent to which Europe should participate in NASA's post-Apollo space programme.

However, the 1968 European Space Conference did take a number of decisions which would ultimately overcome the problems hampering ELDO and ESRO. It authorised an increase in spending and the initiation of projects which would run beyond the original eight-year ESRO planning horizon. Most importantly, it began the discussions on the structural re-organisation of European space activities that would ultimately lead to the creation of a single European Space Agency which would absorb ESRO, ELDO and CETS (the European Conference on Satellite Communications), and which would be responsible for the launching of applications satellites as well as for pure scientific research.[38]

In 1971 the ESRO states agreed to include the development and launching of applications satellites within ESRO's remit. This decision broke the logjam of disputes holding up progress and in December 1972 unanimous agreement was reached on all outstanding issues. Three major programmes were proposed in order to overcome the disputes between Britain, France and West Germany. A new launcher (Ariane) would be developed by France; West Germany would lead the development of *Spacelab* as the European contribution to the post-Apollo NASA programme, and Britain would produce MAROTS, a maritime communications applications satellite.

France provided 62.5 per cent of the total funding for Ariane, West Germany 52.5 per cent of the *Spacelab* funding, and Britain contributed

56 per cent of the funding for MAROTS. In addition, each of the countries agreed to make some financial contribution to the other two projects.

The European Space Agency (ESA)

The European Space Agency came into formal existence in May 1975, with the signing of the ESA Convention by all the ESRO and ELDO states except Australia. The new organisation defined its purpose as being 'to provide and to promote, for exclusively peaceful purposes, cooperation among European states in space research and technology and their space applications, with a view to their being used for scientific purposes and for operational space applications systems'.[39]

As Zabusky has noted, this statement of the Agency's purpose reveals 'that the work of ESA is not in the execution of research and development in space science and technology, rather the work of ESA is cooperation'.[40] ESA's policies are a functionalist effort to concretise the practice of *unity in diversity* that has characterised the European integration project since the early 1950s. Although the work of ESA represents a significant contribution to the European integration process, the Agency itself is not engaged in a process of integration as such. Rather, its purpose is the *harmonisation* of European policies, so as to avoid unnecessary overlap or duplication of effort, while making possible larger-scale projects that would be beyond the resources of any single state. Its approach to harmonisation recognises the divergence of national interests and the importance of conceptions of national sovereignty in Europe, and it does not attempt to control or direct all European space activities.[41]

As with ELDO/ESRO, the new European space organisation was clearly perceived as being, among other things, a vehicle for the political unification of Europe via functional integration. In 1984, on the twentieth anniversary of the foundation of ESRO, the ESA Director General declared that, 'The European space effort is an outstanding demonstration of what we can do on the old Continent when united; let it be an example for European unity in a broad sense'.[42]

By the time that the existing European space organisations were merged into ESA, the neo-functionalist ideology of European integration had begun to gather considerable momentum, so that within European institutions, integration had begun to be asserted as a value in itself.[43] Certainly ESA's self-perception incorporated this outlook. An official history described the Agency's staff as seeing themselves as belonging to a European unity and of the Agency itself as 'one of the melting pots for the material from which Europe is gradually being forged, and in which nationalist preoccupations have to give way to wider, more promising visions'.[44] The Report goes on to declare that 'the European Space Agency then, represents a tool for working towards a united Europe – as it were a "European Space Community"'.[45]

However, for the member states, the logic of autonomy could work in different ways, so that in practice harmonisation has made it possible for European states to avoid dependence on the United States, without necessarily creating a unified European programme at the expense of national programmes.[46] National programmes have remained of crucial importance to European governments, particularly in the field of applications satellites.

ESA had the advantage of being able to learn from the mistakes of ESRO and ELDO and could go on to create a more coherent and balanced programme than had previously been achieved.[47] It can be argued that at the beginning of the 1960s Europe was not ready for a completely unified approach to its space programme, as was seen in the decision to create two separate space organisations, ESRO and ELDO.[48] The desire for a merger of all the existing European space bodies in the early 1970s therefore illustrates that European space science and technology had significantly matured and with a decade of cooperative experience within the space organisations and within the EEC itself, the political views of the European states had also matured and Europe was now capable of acting as an entity with a clear political will and consistent policy.[49] ESA absorbed the policy functions of ELDO, ESRO and CETS and, in addition, the ESC was abolished, with its role being taken over by the new ESA Council meeting at Ministerial level.[50]

In order to overcome the programmatic problems that had hampered ELDO and ESRO, the new European Space Agency created a two-tier structure of finance and involvement. A basic core programme of scientific exploration was established. This, along with the administrative costs of running the organisation, was funded from a general budget to which all member states would contribute in proportion to their GNP. All other programmes would be optional. These would include launcher and satellite development, and member states would be free to decide whether or not to participate in these elements and to some extent, at what financial level. Only those states participating in an optional programme would have any subsequent say in the programme's planning or development.

While the creation of ESA did not immediately overcome all the problems that had dogged its predecessors, it certainly resolved some, and reduced the impact of others. The emergence of ESA clearly represents a turning point in the history of Europe in space. The new ESA Convention laid particular stress on positive internationalism within the European context. As well as the Agency's general function in supporting cooperative efforts, the Convention obliges member states to offer cooperation within the overall ESA framework on space projects which they initiate. In addition, Article 8 insists that ESA, in procuring launch facilities, will give preference to those 'developed within the framework of its programmes, or by a member state, or with significant agency contribution'.

The functions of ESA are numerous, and include defining and implementing long-term European space policy, recommending space objectives to the member states, and harmonising the space policies of the member states. This involves coordinating the overall European space programme with the separate national programmes, and integrating the latter as completely as possible in the European space programme, particularly with respect to the development of application satellites. In addition, ESA implements collaborative programmes and activities in the space field and is responsible for defining and implementing a coherent industrial policy appropriate to its programmes and serves as a contracting agency for industrial contributions.[51]

The manner in which ESA sub-contracts work also makes an important contribution to the process of European integration. In contrast to ELDO and ESRO, instead of giving contracts to individual companies, ESA policy is to give the contracts to consortia which are created out of national companies from several member states. This 'indirect intervention' policy favours the formation of multinational industrial networks such as MESH (Matra, ERNO, Saab, British Aerospace) or STAR (British Aerospace, Dornier, AEG-Telefunken, VFW).[52] ESA has attempted to promote European consortia not only in order to make European industry cost-effective in achieving ESA's goals, but also to make it competitive by international standards.[53]

Even on programmes with a dominant national actor, sub-contracting of key system elements produces a multinational effect. Thus, for example, while France contributed 62.5 per cent of the funding for the Ariane launcher,[54] key components were provided by non-French companies. These included Contraves (Switzerland), for the payload fairing, Messerschmidt-Boelkow-Blohm (West Germany), for the second stage and liquid-fuel boosters, SNIA-BPD (Italy), for the solid rocket boosters and British Aerospace for the Spelda multi-payload deployment system.[55] Overall, ESA has played an important role in the development of European technology and the success of ESA in hardware development is considered one of the best examples of what European integration can achieve.[56] The international competitiveness of European space industry increased during the 1990s, as the major companies began a series of mergers designed to keep them competitive in the face of similar developments in the post-Cold War American aerospace industry.

ESA departed from its predecessors by developing long-term planning which looked up to 20 years ahead and committed the organisation to a mix of large and smaller projects spanning the scientific disciplines.[57] This planning structure emerged from an initiative of the scientific and industrial lobbies rather than from national governmental level.

The Ariane programme was crucial to the subsequent success of ESA, both in enabling it to escape from the sense of failure that the Europa legacy had left, and in terms of giving Europe access to the growing market for launching applications satellites. The first Ariane was successfully launched on 24 December 1979, and the rocket went on to become an outstanding technological and commercial success in the 1980s and 1990s.

The European Union and space policy

'European space policy' is not confined to ESA but also embraces the European Union, which has always seen itself as the keystone organisation for the integration of Europe. The process of European integration began in 1951 with the creation of the European Coal and Steel Community, whose founding Treaty of Paris referred to its objective to 'create, by establishing an economic community, the basis for a broader and deeper community among peoples long divided by bloody conflicts'.[58] It was crucially significant that this initial effort to begin integration focussed on the key industries that had underpinned French and German war-making capacity over the previous century.

During the period from the early 1960s to the end of the Cold War, ESRO/ELDO and ESA were in a sense the space dimension of a broader pattern of European integration and international cooperation that was demonstrated in other fields in organisations such as NATO, the European Community and Euratom. However, it was not institutionally linked in any ways to these other bodies, and its membership only partially overlapped with those of the other organisations. Nor did the other organisations concede ESA any *exclusive* right to dominate developments in this area of activity. NATO was involved in space activities through the NATO series of communication satellites, while the EU believed it had competence in the space policy realm in areas that were relevant to its other integrational activities, such as industrial policy in the aerospace sector, under articles 70, 154, 157 and 163–73 of the Treaty of the European Communities.

The idea that the European Community provided a logical framework within which *all* European integration efforts should take place had been suggested as early as 1966 by the Western European Union. The WEU Assembly called upon its member governments to 'prepare for the inclusion of a permanent European space vehicle launcher development organisation within the framework of a future single European Community'.[59]

The formal involvement of the European Community with space policy may be said to have begun with the Commission's participation in the 1970 European Space Conference, when it was invited to attend as an observer. From that point onwards it became traditional for the Commission to attend ministerial meetings of the ESA Council as an observer.[60]

After the formation of ESA, the European Community recognised that ESA was the *lead* European organisation in the exploitation of space and the integration of European space policy.[61] However, it was recognised that this might change in the longer term, both because the Community was a major potential user of space systems, and because the history of the European Community since 1957 had been characterised by a steady increase in the number of areas affected by the integration process.[62]

With the signing of the Single European Act in 1987, the EU possessed an interest in space for four reasons: as a focus for developing certain types of applied space technology; as a sponsor of research and development programmes, some of which have a space element; as a key force in shaping the market conditions in which the space industry operates; and as an increasingly active actor in the security field.[63] In 1989 the EC and ESA established five joint committees, charged with exchanging information and perspectives on matters of common interest. This resulted from a meeting between the Commission President, Jacques Delors and the Director-General of ESA, Reimar Lust, who had agreed on a 'declaration of solidarity concerning future cooperative relations which would be instituted on the basis of respect for one another's competences'.[64]

The Single European Act added research and development and environmental policy to the Community's competences and 'the potential scope of this remit necessarily included the space field in general and remote-sensing in particular'.[65] The Commission quickly produced a position document on space issues which surveyed the strengths and weaknesses of existing arrangements for space activities by Europe, and particularly noted the absence of a proper framework for cooperative space activities in the defence area.[66]

In 1991 the European Parliament produced a report which noted, among other things, that while ESA was not politically equipped to 'enforce an overall European space policy', the EU in contrast possessed a huge range of legal and constitutional mechanisms designed to allow it to define and implement common policies, and these mechanisms clearly gave the Community the opportunity 'to play a crucial role in making it possible for Europe to reap the full benefits of ESA's outstanding achievements'.[67] In 1992 the Space Advisory Group was created as a discussion forum of high level representatives from ESA, the EU Commission and the national governments of the member states. Meeting twice a year, the SAG's primary function is to identify areas where coordination is necessary.[68]

In the 1990s, the European Union increasingly argued that the Union needed its own comprehensive space policy, given the ubiquitous and accelerating importance of space technology in many areas affecting the well-being of the EU's citizens. In 2003 the European Commission published a White Paper on European Space Policy which argued that it was time to place space policy issues 'on the Union's political agenda at the heart of the European construction process by putting space applications linked to inspirational goals at the service of the enlarged Europe and of its citizens'.[69]

The White Paper emphasised the ability of space applications 'to satisfy the needs of the citizens and to respond to the Union's political objectives' and therefore called for statute changes to give the EU 'new responsibilities for driving, funding and coordinating activities within an extended space policy'.[70] This clearly had major implications, both for the direction in which the Commission was seeking to take the Union, and for the relationship between the EU and ESA.

To a large extent the EU had already crossed this threshold. In February 1999 the European Commission proposed the development of a European satellite radio navigation system, Galileo. In May 2003 the ESA and EU authorised the implementation phase for development of the Galileo constellation of 30 satellites and associated ground stations. Galileo represents the first major space project organised under the aegis of the EU, and the first time that the EU would directly control a strategic space asset of this sort. The European debate over the merits of the Galileo proposal also firmed up the concept of 'European non-dependence' that underlay EU thinking on space policy.[71] European users already had access to the American GPS system, but concerns that such access would be limited or blocked during American military campaigns triggered a similar response to the move into launcher technology in the 1960s. It was not enough to have guaranteed access to US capabilities virtually all the time, Europe needed a system under its own control that it would have access to on a permanent basis.

The international security dimension

The Galileo debate was triggered by the impact of the military uses of satellite technology. But Galileo itself was symptomatic of European limitations in this particular field. Although a debate on the military potential of Galileo for European states began after the NATO Kosovo campaign in 1999, neither the EU nor ESA were able to resolve the issue of how Galileo might be used in time of conflict.[72]

A fundamental aspect of ESA's activities and those of its ESRO predecessor has been a commitment to pursue only the 'peaceful' utilisation of space. This limitation is enshrined in the ESA Convention. However, the Convention does not define what 'peaceful purposes' are, or are not, in this context, leaving scope for dispute and a degree of permissive interpretation.

The move to develop a coherent European space policy has been located clearly within the expanded security paradigm. During the final years of the Cold War, the definition of security used by the international community began to expand beyond the traditional military understanding to encompass other areas of human vulnerability. These included economic and environmental threats and threats to societies. The broader definition of security enabled the EU to pursue its goal of acquiring a 'security' identity, without simply duplicating the competences of NATO. It allowed for the use of military assets in a security-building role, particularly when weapons are not involved, for example in the use of reconnaissance satellites for confidence building, and verification of compliance with arms control agreements. Madders refers to this as the doctrine of 'legitimate peaceful security activities'.

Historically, NATO had been the key organisation providing cooperative European military security. The EU's competence had been in areas such as economic development, and latterly, coordination of environmental protection policies. But as the definition of 'security' expanded to embrace areas such as the economy and the environment, the EU found itself with legitimate security roles, making it far easier to move one stage further and enter the military security arena. The importance of European space capabilities in this regard is that satellite systems clearly have an enormous amount to contribute to the achievement of security objectives in the economic, environmental and societal areas, as well as in more traditional military areas. According to the European Commission, space technologies can make valuable contributions to economic growth, job creation and industrial competitiveness; a successful enlargement of the Union; sustainable development; a stronger security and defence for all; fighting poverty and aiding development'.[73] All of these can be seen as being embraced within a broader definition of security than the traditional narrow military one.[74] The 2003 *Green Paper* consultation document produced by the Commission and ESA refers to this interpretation of security, 'taking into account the complete spectrum of security-related activities'.[75] The objective for the EU would be to 'raise the EU's political standing in the world and 'help Europe to be a better neighbour and a respected global partner'.[76]

The EU has stressed that to achieve an effective common foreign and security policy, the Union must have access to reliable autonomous intelligence-gathering capabilities, and that in this regard space capabilities are essential. The Union therefore requires 'the capacity to launch, develop and operate satellites providing global communications, positioning and observation systems'.[77] At the Gothenburg EU Summit in 2001, the European Council called for establishing by 2008 a European capacity for global monitoring of the environment and security. In response, the European Commission and ESA initiated the Global Monitoring for the Environment and Security Programme (GMES). GMES is designed to support sustainable development policies in areas such as the environment, agriculture, fisheries, transport and regional development. More specifically, it will provide support for objectives 'linked to the implementation of a Common Foreign and Security Policy as well as to allow early warning and rapid damage assessment in natural disasters'.[78]

David Mitrany, one of the 'fathers' of European integration, felt that in the contemporary era of 'satellites and space travel, we have in truth reached the "no man's land" of sovereignty'.[79] The European space programme is distinctive in that, while it serves political purposes as important as those of the American, Chinese or Indian programmes, in the European case the ideological purpose is multinational and integrationist, rather than national. While the Europeans' struggled to operationalise their space goals during the 1960s, their difficulties simply mirrored those being experienced in the economic and military dimensions of the European integration project.

Since the 1970s ESA has been spectacularly successful, both in terms of the completion of numerous ambitious space exploration projects, such as the missions to Halley's Comet, Mars and Saturn, and in regard to the commercial success of the Ariane satellite launcher series. It is a testament to its technological, economic and political success both that it should find itself being courted by the European Union, and that it should be facing the dilemma of entering the realm of military space activities at the start of the twenty-first century.

Chapter 6

Space as a military force multiplier

> But I ask the question: short of a revolution in the heart of man and the nature of states, by what miracle could interplanetary space be preserved from military use?[1]

Military uses of space

The military uses of space have been discussed so far throughout this book in relation to the space programmes of particular countries. The purpose of this chapter is to focus on the idea of space as a military force multiplier and to look at two important distinctions, those between *passive* and *active* military space systems, and between the *militarisation* and *weaponisation* of space.

These distinctions can be related also to those that differentiate the 'high ground' and 'theatre' advocates of space militarisation. For the former, space is simply a medium in which operations occur that significantly enhance the effectiveness of terrestrial forces. For the latter, space is a 'mission', as well as an environment and needs to be defended and exploited in its own right.

Military space systems had become an important part of superpower military operations as early as 1963.[2] During the Cold War, the superpowers deployed military satellites as passive force multipliers. Force multipliers are military systems that significantly increase the effectiveness of combat forces. Space systems have particularly multiplied the effectiveness of military forces by providing them with far greater intelligence information about enemy force dispositions and, in recent decades, allowing real-time imagery of such forces. A passive military space system can be defined as one that 'is not in itself a weapon, but can be used to support military activity'.[3] Systems of this sort can be grouped into five categories: reconnaissance, communication, geodesy, meteorological and navigation satellites; the use of these increase the effectiveness of terrestrial military forces.[4] Many of these categories can themselves be further subdivided; for example, routine reconnaissance and early-warning satellites, or photographic reconnaissance versus electronic intelligence satellites. Such systems became increasingly important to the superpowers during the Cold War, so that 'of the 3174 satellites orbited between 1957 and 1985, some 75 percent were launched to enhance the performance of nuclear and other weapons on Earth'.[5] By the end of the twentieth century such systems had become central to the military operations of the major powers, so that, for example, 'practically every piece of information used by the US military today is either derived from or transmitted through space'.[6]

Different countries may use different procedures to achieve the same operational capability. With regard to reconnaissance satellites, for example, the United States has tended to launch a small number of large, highly effective satellites, which remain in orbit for long periods. The Soviet Union and subsequently the Russian Federation, in contrast, have tended to launch satellites with much shorter periods in orbit, and have therefore needed to launch much larger numbers in order to gain the same degree of global coverage.[7] Reconnaissance satellites operate in low orbits, and are quickly affected by atmospheric drag from the upper reaches of the atmosphere. American satellites manoeuvre in orbit, allowing them to be periodically boosted to a slightly higher orbit and thereby extending their time in space.[8] The Soviet Union also orbited military space stations for reconnaissance purposes, such as Salyut-3 launched in June 1974.[9]

Over time, the number of military functions to which satellite systems make an important contribution has increased enormously. US meteorological satellites, for example, are used not simply to provide general weather forecasts for conflict zones, but to give detailed information relevant to a wide range of decisions; for example those relating to 'resource protection, operational timing, flight planning, ship routing, munitions selection, chemical attack dispersion predictions, radar and communications anomaly resolution and targeting'.[10]

Military communications satellites in turn have become more and more important in terms of managing the battlespace and improving situational awareness. They are particularly valuable in the 'expeditionary wars' fought in remote and difficult terrain, where there may not be an existing well-developed communications infrastructure, that have become typical of the asymmetric conflicts waged since the end of the Cold War.[11] The most dramatic change in military satellite usage in the post-Cold War period has been the reliance on satellites at the tactical level, rather than simply the strategic level, as was largely the case previously.

Military space technology often exhibits a Janus-face which makes it difficult to condemn or condone simplistically. Successive US space policy statements have reflected this. Thus, for example, the 1982 Reagan Administration Space Policy statement declared that 'the United States is committed to the exploration and use of space by all nations for peaceful purposes and for the benefit of mankind. "Peaceful purposes" allow activities in pursuit of national security goals'.[12]

Reconnaissance satellites, for example, perform a number of functions, including ISTAR (intelligence, surveillance, target acquisition and reconnaissance), signals monitoring, photo reconnaissance, infra-red reconnaissance, electronic reconnaissance, ocean reconnaissance and ballistic missile attack early warning. They can gather visual and photographic data, eavesdrop on signals intelligence and collect radar defence information. All these functions are crucial for effective warfighting.

However, the same technologies have other crucial uses. As well as gathering information in preparation for, or on the conduct of war, they can be used for the monitoring of compliance with arms control agreements, and can play a crucial part in underpinning confidence-building regimes. During the Cold War a series of arms control agreements introduced an important element of stability into the east–west relationship. Given the degree of mutual suspicion held by NATO members and the Warsaw Pact countries for one another, it was only the ability to verify compliance that made the two sides willing to sign such agreements.[13] Most of the monitoring was done using satellite systems of various kinds.

Another crucial contribution to Cold War stability was provided by the early warning satellites. These increased the amount of warning time of a nuclear attack available from a few minutes to about half an hour, allowing the superpowers to move away from the 'hair-trigger' alert postures that had been typical in the late 1950s and early 1960s.

An active military space system, 'is either a weapon in its own right, or an inherent part of an overall weapon system'.[14] A satellite placed in orbit and armed with weapons designed to destroy another satellite, or to intercept a nuclear warhead during its ballistic flight through space, would constitute active military space systems. They are weapons themselves, unlike passive systems, that are not weapons, though they may have the effect of making other weapons operate more efficiently.

The definitional differences blur at the margins, and some systems can fit either. For example, a radar system can be viewed simply as a reconnaissance capability and therefore passive, or as a 'gunsight' for targeting, and therefore part of an active system. In 2003, US Under Secretary of the Air Force Peter Teets described the proposed US space-based radar as acting 'as the forward eyes for strike platforms and other intelligence, surveillance, and reconnaissance (ISR) assets by detecting surface movers (ground target indication) and rapidly imaging stationary targets (synthetic aperture radar).[15]

The 1967 Outer Space Treaty forbade the placing of weapons of mass destruction in space, but deployment of conventional weaponry was only banned from 'celestial bodies'. Weapons of mass destruction are those armed with nuclear, chemical or biological warheads. The effect of the treaty was to ban conventional weaponry from places such as the Moon or Mars, but to allow its deployment in Earth orbit or near-Earth space. Notably, the treaty did not restrict the use of space for passive military purposes, including the use of reconnaissance, communications, meteorological and navigation satellites.[16]

The debate over weaponisation

Whereas space has been *militarised* since the advent of the space age, it has not thus far been *weaponised*. It is militarised in the sense that large numbers of satellites have been orbited that perform a primary or completely military function, such as military reconnaissance or communications satellites. The weaponisation of space would require the deployment of 'destructive devices moving under the influence of gravity in near or "planetary" space that can be aimed either against other space objects or against targets on Earth'.[17] To date active devices of this sort have appeared briefly in space during tests of such weapon systems, but have never as yet been operationally deployed on a permanent basis.

Concerns over the potential emplacement of weapons in space are not new, but have resurfaced strongly since the advent of the George W Bush administration, because of the administration's renewed commitment to the emplacement of space-based anti-ballistic missile systems, and its interest in the development of anti-satellite systems. These developments raised concerns among the international community that a new period of space weaponisation might be about to occur. For example, in 2002 the Preparatory Committee of the Nuclear Non-Proliferation Treaty expressed international concerns about the possibility of an arms race in outer space.[18]

The conflicts in Afghanistan and Iraq demonstrated the value of space-based reconnaissance, communications and targeting capabilities to the forces of the United States and its allies. Since Operation Desert Storm in 1991, a pattern has emerged of each subsequent conflict showing a continuing increase in the capabilities represented by US military space assets, and the increasing reliance of US forces on those assets. The Clinton administration recognised this in its national space policy, which emphasised that while the United States was committed to the use of space for peaceful purposes, the term 'peaceful purposes' was deemed to allow defence and intelligence-related activities in pursuit of national security and other goals'.[19] In 1999 the national security strategy, announced by the Clinton administration, for the first time declared that unimpeded access to and use of space was *a vital national interest of the United States.*[20]

An increasing number of senior American military officers and space commentators have argued that the weaponisation of space is an inevitable development, that in the future effective military campaigns must include offensive space operations against adversaries that possess space-based force multipliers. This will mean a 'transition from using space assets to support combat operations on the surface of the earth to using space assets to conduct combat operations in space, from space and through space'.[21] Senator Bob Smith, a leading Congressional advocate of space weaponisation, argues dramatically that without it 'we will become vulnerable beyond our worst fears'.[22]

The argument has also been put forward that throughout history there has been a connection between the way societies make wealth and the way that they make war.[23] As the commercialisation of space gathers pace, therefore, it would be only a matter of time before space in turn became another medium for warfare. This creates a requirement for those states that are currently most dependent on space commercially and militarily, to begin preparing for the inevitable conflict in that medium. The United States, the state that has become both the most powerful in space and the most dependent on it, inevitably feels pressured to take the lead in such a development.

In future major conflicts, the US needs to be able to exploit its asymmetrical advantages in information warfare, to protect its capability to do so, and to deny those same capabilities to the adversary.[24] The US military increasingly believes that weapons will be required to deal with space-related conflict.[25] Secretary of Defense Rumsfeld stated in 2001 that the overall policy remained what it had been since 1996, that 'consistent with treaty obligations, the United States will develop, operate and maintain space control capabilities to ensure freedom of action in space, and if directed, deny such freedom of action to adversaries'.[26]

Historically, much of the opposition to the deployment of space weaponry has derived from concerns about the effect of such a development on the ABM Treaty of 1972. With the withdrawal of the United States from the treaty, this factor is no longer relevant as a restraining factor on space weaponisation. The only treaty that specifically forbids such weaponisation is now the Outer Space Treaty of 1967, which relates specifically and solely to weapons of mass destruction. There is no international treaty obligation for the United States to refrain from deploying conventionally armed weaponry in space. US Space Command was forthright in 1998 in arguing that 'early in the 21st Century, space will become another medium of warfare'.[27] Whether the United States will actually cross this threshold, and if it does, what time-scale it might follow, are issues which will be determined by the American cultural and political systems, rather than by external pressures or international obligations.

For the strongest proponents, the logic is inexorable. The United States should act as soon as possible to achieve total military dominance of low-Earth orbit. 'From that high-ground vantage, near the top of the Earth's gravity well, space-based laser or kinetic energy weapons could prevent any other state from deploying assets there, and could most effectively engage and destroy terrestrial enemy ASAT facilities'.[28]

A wide range of elements would need to be in place if the United States were to proceed to achieve space control, but three 'stand out as especially critical: sound doctrine, viable technology and political will'.[29] Such debate as there has been on this subject tends to focus on the difficulties or otherwise of acquiring the technology, but it is the doctrinal and political criteria that are the most crucial, and ultimately determining factors.

Historical advocacy of space weaponisation

At the dawn of the space age the Eisenhower administration came to the conclusion that the weaponisation of space was not in the interests of the United States.[30] There was a difference in outlook, however, between the political leadership and the uniformed military on the issue. Air Force leaders in the late 1950s saw space as a potential area of military activity, where the United States had an opportunity to gain pre-eminence over the Soviet Union. The space age began with the launch of Sputnik I by the Soviet Union on 4 October 1957. The United States became a participant with the launch of Explorer I on 31 January 1958. As noted in Chapter 3, the launch of Sputnik profoundly unsettled the United States. General Thomas D. White, the Air Force Chief of Staff, declared that 'in the future, whoever has the capacity to control space will likewise possess the capability to exert control over the surface of the earth'.[31]

However, the American political élite was far more cautious about the desirability of placing weapons in space. Four days after the launch of Sputnik, in October 1957, US Deputy Secretary of Defense Quarles was briefed by senior Air Force commanders on the potential of reconnaissance satellites and of satellite offensive operations against Soviet spacecraft. Quarles approved the reconnaissance programme, but insisted that placing weapons in space was against both the policy and the interests of the United States.[32]

Similar attitudes existed in Congress. The Chairman of the House Committee on Science and Astronautics noted that the US had both political and utilitarian grounds for opposing the weaponisation of space. Since the US was virtually impervious to attack, apart from threats such as ballistic missiles moving through space, 'we have not only the propaganda value, but we also have a very strong nationalistic reason for wishing to make outer space peaceful'.[33]

Political attitudes towards the security uses of space have in fact frequently been at odds with the views of the military. During the Cold War reconnaissance satellites played a major and crucial role in maintaining strategic stability and allowing the conclusion of arms control agreements. Yet General Gavin had argued in 1958 that it was 'inconceivable', that we would indefinitely tolerate Soviet reconnaissance of the United States without protest'.[34] In fact the United States not only tolerated such reconnaissance, but welcomed the Soviet initiative, because it allowed the United States to reciprocate and gained far more from doing so than did the Soviet Union. In

1967 President Johnson claimed that the information the United States had gained from its military reconnaissance satellites was worth ten times the money that the US had invested in its space programme.[35]

The temptation to place weapons in space was not always confined to the military, however. Following the emergence of significant Soviet strategic nuclear capability in the early 1960s, the US State Department suggested that 'we should develop as rapidly as possible anti-satellite capabilities'.[36] With the election of the more space minded Kennedy-Johnson administration in 1960, the military hoped for a more receptive attitude towards the militarisation of space. However, although Kennedy dramatically increased the overall scale and political support for American space activities, he did not alter the policy inherited from the Eisenhower administration of supporting space for 'peaceful purposes'. Indeed Kennedy cancelled the Satellite Interceptor (SAINT) development programme for fear that it might generate an offensive military space race with the Soviet Union. Ground-launched missiles capable of attacking satellites were however deployed at bases in the Pacific, but were decommissioned in the early 1970s.

However, as US reliance on military space developed during the 1960s, this assumption began to be questioned within the USAF. In 1968 General Oris Johnson noted the military dominance of the Soviet space programme and suggested that 'the necessity for effective space defense weapons is both obvious and urgent'.[37] Studies carried out by the Ford administration as early as 1976 suggested that the United States was becoming increasingly dependent on satellites for various functions and that little provision had been made for satellite survival in wartime.[38] Two days before he left office in 1976, President Ford ordered the rapid development of an American anti-satellite weapon.

Under President Carter's Presidential Directive 37, the United States adopted a policy of rather more openness about its use of space for passive force enhancement purposes. In 1977 the USAF adopted three policy implementation responsibilities, which included maintaining the freedom of space via its space capabilities and 'military operations in space, conducted by the letter and spirit of existing treaties and in accordance with international law'.[39] The Carter administration's preference was for neither superpower to deploy ASAT systems, but rather to negotiate an arms control regime that would constrain ASAT technologies.[40]

The first US administration to openly advocate the weaponisation of space was that of President Reagan. In 1982 the revised US National Space Policy called for deployment of an operational anti-satellite system. A system was developed in which high-flying F-15 aircraft were armed with heat-seeking missiles that would be launched at maximum altitude as the target satellite was passing overhead. The Reagan administration also promoted the idea of space-based weaponry as part of the architecture of the strategic defence initiative announced in March 1983, to develop a system capable of protecting the United States against incoming ballistic missiles.

The attitude of the Clinton administration was somewhat ambiguous. The national space policy under Clinton directed the Department of Defense to 'maintain the capability to execute the mission areas of space support, force enhancement, space control and force application'. But while 'force application' thereby became a recognised part of national space policy, the administration made no effort to pursue or deploy the technologies that would have given such a policy meaning.

Iraq as the first 'space war'

Space power first came to dramatic prominence, both within and outside the American armed forces, at the time of Operation Desert Storm, the First Gulf War, in 1991. Satellites had played a role in earlier conflicts, for example producing photo-reconnaissance and weather data during the Vietnam War,[41] and photo and signals intelligence during the Falklands War.[42] Under the Reagan administration it was recognised that 'even in a very limited war, we would have an absolutely critical dependence on space today'.[43]

The role of space power was not central to the allied victory in the Gulf War. The official US Air Force history of the conflict only mentions the contribution of space assets on a single page.[44] Nevertheless, a variety of space assets contributed to the success of the operation. The Global Positioning System was not fully operational in 1991, but was nevertheless used on a large scale.[45] Weather data was provided by six Defense Meteorological Satellites. Secure communications were provided by the Defense Satellite Communication system. Satellites provided nearly 90 per cent of all intra-theatre and inter-theatre communications.[46] Scud missile launches were monitored using Defense Support Programme satellites. In addition, two civilian satellite systems, the US LANDSAT and French SPOT were used to provide supplementary imagery. Overall, eight American and allied civilian satellites contributed to the effectiveness of US military operations during Operation Desert Storm.[47] This pattern came to be seen as the norm for future large-scale US military operations, with the head of US Space Command arguing by 1996 that 'because of expanding demands for support, we expect a blend of military, civil, commercial, and international systems to meet our future satellite communications needs'.[48]

When Iraq invaded Kuwait in 1990, it had been purchasing SPOT images for strategic reconnaissance. Soon after the invasion, France embargoed the supply of SPOT imagery to Iraq. Satellite images however proved enormously valuable to the coalition forces. They were used to provide precise maps of the region, to determine suitable areas for land, sea and air assault, to verify target coordinates and to rescue downed pilots. The images made possible very accurate targeting. They allowed a single-building strike against Iraqi General officers in Kuwait City, and accurate attacks on the Iraqi air operations centre, intelligence centre and Ministry of Defence. In addition, by helping to calculate attack angles so that bombs or missiles that might land long or short had the least chance of causing collateral damage, it made it possible to try and avoid hospitals, schools, mosques and residential areas.[49]

During the Gulf War space systems provided a variety of important services. The Defense Satellite Communications System satellites provided high capacity and secure communications. So much communication was routed via satellite that the military systems became overburdened, and civilian systems were also brought into use, for relaying communications that did not need to be secure or which were less urgent.

The NAVSTAR global positioning satellite system, though not fully operational in 1991, also proved extremely valuable, particularly given the featureless nature of the terrain on which much of the military activity took place. Surveillance was also crucial, and satellites provided optical, radar and infra-red images for coalition commanders. This was valuable for intelligence gathering, battle-damage assessment and ballistic missile launch warning. Electronic intelligence gathering was also important.

Desert Shield/Desert Storm also showed up important weaknesses in the US exploitation of space power at the time. For example, Desert Storm forces were at first unable to obtain timely weather and imagery intelligence, because shipment of the necessary user equipment to the military theatre was not given a high priority in the planning for the logistics build-up.[50] Similarly, the first warnings of an Iraqi SCUD missile strike were received after the missile had hit its target zone, because at the start of the conflict, US space-based missile early warning systems were still following their Cold War procedures of passing information to strategic headquarters in the continental United States, rather than to commanders and troops in the military theatre.[51]

For these reasons, it has been argued that in the Gulf War, the space systems 'did not come close to achieving their full potential'.[52] Even so, Desert Storm was in many ways 'a watershed event for military space applications, because for the first time space systems were both integral to the conflict and critical to the outcome of the war'.[53] During the Cold War satellite systems were the crucial 'fourth leg' of the strategic nuclear deterrent, because deterrence depended in part on the ability of the United States armed forces and national command authority to collect, process and disseminate information.[54] The Gulf War witnessed a crucial transition from the use of space power in support of strategic deterrence, to its use in support of tactical warfighting tasks. Space resources were used at almost every level of the war effort. These developments came only five years after a former Director of the Pentagon's Defense Advanced Research Projects Agency had confidently declared that satellite systems were 'generally viewed by military field commanders as peacetime systems; nice to have, but not to be relied on in wartime'.[55]

The use of space systems was not decisive in Desert Storm; coalition forces would have defeated Iraq anyway. Nevertheless, without the contribution made by space forces, the coalition victory would have been more difficult and more costly in terms of lives and *matériel*.[56] Significantly, the US control of space was not contested by Iraq during the Gulf War, but US planners have to assume that America is likely at some point to find itself in conflict with an opponent possessing significant space assets, perhaps including some ASAT capability. Vice-Admiral William Dougherty included among the lessons of Desert Storm a requirement for a responsive space launch capability and an ability to protect US space assets and selectively deny adversary space use.[57] To an extent, therefore, Desert Storm saw the United States begin 'moving from a sanctuary doctrine towards a survivability doctrine'.[58] Space was no longer seen simply as a dimension where passive military operations could take place, but as a theatre where active combat operations would occur, and for which the United States would need to be prepared.

Active space systems

Active space systems currently fall into two categories, ballistic missile defence, and anti-satellite systems.

As early as the Second World War German scientists had begun work on a follow-on missile to the V-2 which would be intercontinental in range and targeted against New York.[59] The war ended before these plans could be implemented, but in the Cold War that followed both superpowers soon turned to the idea of using such intercontinental ballistic missiles to carry their nuclear weapons to target without fear of interception. This consideration was particularly powerful for the Soviet Union, which unlike the United States, lacked a large long-range bomber force, but the United States in turn knew that in time its manned bombers would become vulnerable to interception by improving Soviet air defences and that its own ballistic missile deployments were therefore only a question of time.

The defensive systems advocated during the early 1960s were extremely ambitious and required technologies that were simply not available, nor would they be for decades to come. US proposals involved hit-to-kill techniques requiring direct hits on the target missile, and carrying out this interception in the target missile's first two minutes of flight. Space-based interceptors of this sort were given the collective title of BAMBI (ballistic missile boost intercept).[60] In the mid-1960s the United States began developing the *Sentinel* ABM system, which would have used ground launched, nuclear armed missiles, to defend America's cities. In 1969 the Nixon administration renamed it Safeguard, and changed its purpose to that of protecting a limited number of American ICBM sites. Only one such base finally gained such protection (Grand Forks, North Dakota), but the ABM system was only operational for nine months before being closed down.

The Soviet Union, in contrast, developed the *Galosh* system of four 16-missile sites deployed around Moscow. This system remained operational throughout the rest of the Cold War and was upgraded with better missiles and radars during the 1980s. The *Galosh* system also had a clear ASAT capability. In addition, the USSR deployed SA-5 missiles with a limited ABM capability around a number of other Soviet cities.[61]

In 1972 the United States and the Soviet Union signed the Anti-ballistic Missile Treaty, part of the overall SALT I agreement. Under the ABM treaty each side agreed to defend only two sites in its territory. A 1974 amendment to the treaty reduced the permitted defensive sites to one, the USSR choosing to defend Moscow and the USA to protect the Grand Forks ICBM launch site in North Dakota. Thus each side was leaving most of its territory and population completely vulnerable to a nuclear strike by the other side. The terrible logic behind this strategy was that with its people so totally unprotected neither side would ever have any incentive to attack, and peace and strategic nuclear stability would be assured. Space-based defensive systems were specifically banned by the treaty.[62] The banning of meaningful defences also opened the way to significant reductions in the number of strategic nuclear weapons each side possessed, since the numbers retained to swamp the opposing defences could now be eliminated.

These considerations and the mutually agreed nuclear deterrence logic that seemed to underpin them made it all the more dramatic when, in March 1983, President Ronald Reagan revived and accelerated the American research programme into ballistic missile defences.

The Reagan BMD research programme was more ambitious than any of those that preceded or followed it. In March 1983 the President challenged America's scientists to develop defensive technology that would render nuclear weapons 'impotent and obsolete'.[63] Because the early proposals envisaged placing most of the defences in orbit, and using exotic technologies such as lasers and particle beams, it was hardly surprising that the programme was quickly dubbed 'Star Wars' by the media and the public. On 25 March 1983, Reagan signed National Security Decision Directive 85, calling for 'an intensive effort to define a long term research and development programme aimed at the ultimate goal of eliminating the threat posed by ballistic nuclear missiles'.[64] The Reagan SDI scheme was an ambitious attempt to protect the entire population of the United States from ballistic nuclear missile attack, using a 'layered' defence system that would intercept incoming missiles at a variety of points along their flight trajectory. The intensive research effort during the 1980s and in the subsequent administration of President George H.W. Bush, was, however, unable to overcome the enormous technological obstacles to creating such a defensive system.

The National Missile Defense research programme under the Clinton administration envisaged the use of satellites for early warning, tracking and targeting purposes, but did not include a space-based kill mechanism. It envisaged a primarily ground-based system 'that could provide protection to the entire country, but only against limited attacks'.[65] President Clinton signed the National Missile Defense Act in July 1999. The Act makes it US policy to 'deploy as soon as technologically possible, an effective system capable of defending the territory of the United States against limited ballistic missile attack'.[66] The Clinton programme allowed for the objectives of the system to become more ambitious as the technology evolved. On 1 September 2000, President Clinton announced that he had decided not to authorise deployment of a missile defence system during his presidency, precisely because the technology was not yet sufficiently advanced.

Like the Clinton administration, that of President George W. Bush also pursued a less ambitious and therefore more attainable programme than did the Reagan SDI. However, the Bush administration took the crucial decision to withdraw from the 1972 ABM treaty so that the research programme would not be forced to remain within the technological limitations that the treaty imposed. Because of the demise of the Soviet Union and American perception of the dangers from 'rogue' states such as North Korea and Iran, the new programme concentrated on providing a degree of protection for the United States and its allies against limited nuclear strikes, such as the accidental launch of a Russian missile, or a small-scale attack from North Korea.

Like its immediate predecessor, the system relies on ground-based weapons and uses satellites for early-warning, tracking, battle-management and communication. However, the programme allows for the possibility of space-based weapons being added to the 'layered defence' in future years as such technologies mature. The satellites that do form part of the system are essential to its effective performance.[67]

Anti-satellite systems are a natural outcome of the growth in the military utility of orbiting satellites. In the First World War, military aircraft were initially unarmed and used only for reconnaissance. However, once the effectiveness of such reconnaissance was clearly demonstrated, armed forces developed fighter aircraft design to shoot down or deter enemy airborne reconnaissance, and to protect the reconnaissance aircraft of one's own side. There is an obvious analogy with the contemporary situation regarding satellite reconnaissance and, from a military perspective, the need to protect one's own reconnaissance and deny the same facility to the enemy is as crucial in the space age as it was in the early stages of the First World War.

Both superpowers experimented with anti-satellite systems during the Cold War, but the Soviet Union was much more overt in its efforts to develop an effective anti-satellite system during the 1960s and 1970s. In 1964 a new division of the Soviet strategic missile forces was created, the

PKO, with the task of 'destroying the enemy's cosmic means of fighting'.[68] In 1968 the USSR began testing a co-orbital anti-satellite weapon. By the time this first test series ended in 1971 the USSR had demonstrated the ability to place hunter–killer satellites in the vicinity of targets characteristic of communications, meteorological, photo-reconnaissance and electronic-intelligence gathering satellites.[69] A second series of tests, using a modified targeting technology took place between 1976 and 1978. That the Soviet Union attached great strategic importance to these weapons and considered them operational was demonstrated by two significant pieces of evidence. Unlike other space launch systems, the ASAT launchers were kept ready to launch at all times.[70] In addition, when in 1982, the USSR carried out its only full-scale manoeuvre practising all aspects of a global nuclear war against the United States, an ASAT test-flight was part of the manoeuvres, suggesting that ASAT operations were fully integrated into the overall Soviet warfighting plans.[71] The United States reacted to the Soviet ASAT programme both with an equivalent system of its own, and with a programme designed to develop protective countermeasures for its own satellites.

The United States developed a more flexible and effective system in which a high flying F-15 fighter aircraft launched a heat-seeking missile at the target satellite as it passed overhead. Successful tests of the F-15 system were carried out in 1984, before Congress banned further ASAT research. The first Reagan administration publicly committed itself to the acquisition of an American ASAT capability. The administration's 1982 Space Policy Document declared that 'the United States will proceed with development of an anti-satellite (ASAT) capability, with operational deployment as a goal'.[72]

In practice however, neither country had heavily prioritised its ASAT programme and by the mid-1980s the two countries were observing an informal, but effective mutual moratorium on ASAT tests. Opposition to ASAT weapons focussed on the fact that such technology was deeply destabilising. Since the ballistic missile early warning systems, arms control compliance regimes, mutually agreed confidence building measures, routine military reconnaissance, strategic weapons targeting systems, and military communications networks were so heavily dependent on satellite usage, it was obvious that in a crisis, the loss of a satellite, whether by enemy attack, or by technical malfunction, would be likely to trigger immediate military response on the assumption that the other side had already launched the first phase of its attack.

This threat to crisis stability made it all the more alarming that some military space specialists argued, not only that the initial outbreak of conflict in space was not something to be feared, but that it should actually be welcomed! General Thomas Stafford assured a Senate Sub-Committee in 1980 that 'under certain circumstances, space may be viewed as an attractive arena for a show of force. Conflict in space does not violate national boundaries, does not kill people and can provide a visible show of determination at a relatively modest cost'.[73] Ironically, Stafford had commanded the US Apollo spacecraft in 1975, during the Apollo–Soyuz joint mission, the symbolic high point of superpower *détente* in space, but had left NASA after the mission to return to active duty with the US Air Force.[74]

Despite the concerns about ASAT systems, it has not proved possible to conclude an arms control treaty constraining their deployment. One major reason for this is that any definition of an ASAT system invariably covers large numbers of innocent systems as well. Technically, for example, 'any space object capable of changing its orbit is a potential anti-satellite weapon, in view of the risk of collision with any other space object'.[75]

Proponents

Much of the advocacy of space weaponisation reflects a confidence that the United States will be the only country able to deploy the technology needed to exploit this capability for several decades.[76] The disappearance of the threat represented by the former Soviet Union means that the United States no longer has to contemplate scenarios where its own deployments will be mirrored by its opponent, or where US actions might legitimise developments that a potential adversary is in a better position to exploit rapidly.

In addition, it is argued that space power avoids some of the political sensitivities generated by other forms of military and political power. It is ubiquitous, but also unobtrusive, but 'even when the presence of space power is well-known, no laws prohibit it from conducting operations over any spot on Earth'.[77] This is significant in the post-Cold War American hegemony, where Washington's overwhelming dominance creates sensitivities even among allies and deep-seated fears among its enemies, actual and potential.

In such a context, the deployment of space weapons may appear attractive to US leaders, even though their utility is clearly politically circumscribed. Space weapons need not be seen as a panacea, and will probably be limited to specific mission roles, in limited and specific circumstances, where they will be important, but not decisive in themselves.[78] Even while filling fairly specific niches, space-based weapons may be attractive as far as those particular tasks are concerned. For example, space power 'may succeed in coercing some leaders by holding high-value, well-defended targets at risk from a space-based attack that neither puts a pilot in jeopardy, nor requires overflight permissions from any other country'.[79] Force application from space is also proposed, a development which would 'allow the application of force against any target on the face of the earth through space'.[80]

The advocates themselves recognise that the US position of unilateral dominance is unlikely to last long and that 'in the new paradigm, the very weapons that drive it will become threatened by their own kind, and the eternal measure–countermeasure contest will be renewed with new dimensions of technology and tactics'.[81]

It is also argued that the United States needs offensive counterspace weapons as a like-for-like deterrent to reduce the risks of attacks against its own deployed space forces. 'The US currently would suffer the most from losing its space forces, so it is imperative to maintain an ability to retaliate if those forces are attacked. The threat of a decisive US response to attacks may be sufficient to deter an attack'.[82]

Crucial to the emergence of space weaponisation would be the abandonment of the idea that space constitutes a strategic sanctuary. As Gray and Sheldon put it, 'in order for space power to reach its full potential however, space must be recognized as a geographical environment for conflict that is, in a strategic sense, no different from the land, sea, air and the electromagnetic spectrum (EMS)'.[83]

Dangers of weaponisation

As part of the ballistic missile defence research and development programme the United States is developing a Space Based Laser designed to operate in low-Earth orbit to attack missiles during their boost phase. Such a weapon has a clear ASAT potential and the capacity for force projection against targets in the air and on the Earth's surface. The programme director for the system has stated its potential for use against air targets.[84]

Opponents of space weaponisation can be divided into distinctive groups, whose logics are significantly different.[85] Some advocate the continuing maintenance of space as a sanctuary on general arms control principles, others because they see such weapons as potentially profoundly destabilising. There is also a realist approach who argue that it would reduce rather than enhance US security.[86] The logic of this argument is the contention that as the leading space power, the United States has the least to gain, and the most to lose, from encouraging competition in such weaponry.[87]

Bruce DeBlois argues that while there would be obvious military advantages to the United States from being the first country to weaponise space effectively, these benefits need to be balanced against longer-term military disadvantages, 'as well as against broader social, political and economic costs'.[88] Stressing the heavy opportunity costs of investment in space weapons, DeBlois argues that the same tasks could be performed by aircraft, without the political costs of weaponising space.

However, like virtually all American authors, DeBlois, while arguing in favour of arms control and against weapons deployment, supports a policy of 'hedging' against the unforeseen by investing in space weapons research and development. 'Pursuing space-sanctuary policy does not preclude being prepared to do otherwise'.[89] Similarly, Hyten, while advocating a negotiated regime to manage space peacefully, argues that 'as the world's most space-dependent nation, the United States must prepare itself to respond to threats to its national interests should negotiations fail'.[90] Hyten emphasises that 'the United States should use space-based weapons only as a last resort but should not consider such use an unthinkable option'.[91]

ASAT attacks carried out against orbiting satellites might introduce secondary effects, if a highly destructive kill-mechanism was used. While space is an almost limitless domain, in reality the preference for certain orbits over others leads to particular areas being heavily populated by satellites. This grouping together makes the satellites vulnerable to secondary satellite kills as satellites are hit by the debris created by earlier strikes. This would have a dramatic domino effect as subsequent collisions created yet more debris. As the largest user of satellites, the United States would inevitably suffer more than any other if there were a dramatic increase in space debris. Moreover the damaging effects of such a debris cloud might well persist for many years after the end of the conflict.

In the context of a medium- or long-term conflict in which a campaign is waged against an adversary's space power, attacks against targets such as space launch facilities would come into play. These are militarily soft targets. In any major conflict between space powers, launch sites would probably be lost and would take several years to reconstitute, causing significant global economic and political effects in the post-war era. In the longer term, the development of aircraft-like launch systems located at multiple sites might reduce or eliminate this particular vulnerability.

Such a longer term campaign would also move the United States back towards a 'total war' posture in terms of the determination of targets deemed to be legitimate. Attacks would be directed not just against physical assets such as launch sites and orbiting space craft, but also against production facilities, data reception sites and 'the many technicians, operators, analysts and management personnel who create and operate these highly technical systems'.[92]

There is no doubt that US satellite systems do represent a vulnerable centre of gravity for American military power, and indeed of US power more generally. When the Galaxy IV commercial satellite malfunctioned in 1998, there were a number of dramatic effects, including the failure of 35 million pagers, and loss of communication links for several television and radio stations.[93] The US armed forces have become significantly space dependent, but so too have very large areas of commercial and cultural American life become dependent on the continued function of satellite capabilities. There are also synergies between the two. During Operation Iraqi freedom in 2003, over 80 per cent of all US satellite communications used by the US military was provided by commercial satellites.[94]

The growing international space industry means that technology such as reconnaissance and communications capability is now available commercially for countries that do not have their own national military space programmes.

The experience of Desert Storm demonstrated clearly the military potential of commercial communications satellites and Earth observations systems. Commercial companies, such as the SPOT Image Corporation, now openly advertise the military utility of SPOT imagery.[95] Reconnaissance satellites with ground resolutions between 10 and 20 metres can have significant military utility, while resolutions below 10 metres provide extremely valuable military data.[96] The SPOT 5 series has a resolution of around 2.5 metres. The Helios European military reconnaissance satellite has a resolution of 0.3 metres.[97]

Nations with space capabilities can be divided into three tiers. The first consists of 'those states with dedicated military and civilian space capabilities on the cutting edge of technology. Second tier states develop and use dual-purpose space systems for both military and civilian purposes: and third tier nations lease or purchase space capabilities'.[98] Both space- and ground-based attacks on satellites become less politically attractive in the context of a globalised and interdependent world, where the interdependence of space assets means that a spacecraft attack would often affect several different states.[99]

The implication of this is that in wartime, the United States might have to attack commercial satellites belonging to other countries with which it is not at war. There are obvious international political implications raised by this possibility, and also questions raised about the relevance of the current laws of neutrality. Peter Teets, Under Secretary of the Air Force, declares frankly in Counterspace Operations that the United States will need to attack space-borne imagery collectors, commercial or national, that threaten American lives in wartime and specifically identifies the risk of an adversary using the European Galileo positioning system to target American forces in a future conflict.[100]

Whether the attacks would even be militarily worthwhile is debatable. A potential adversary who has recognised the military utility of space reconnaissance is unlikely to wait until war breaks out before acquiring useful imagery. More probably it will have been acquiring relevant imagery from commercial sources for many years beforehand. 'If the enemy has developed a strategic database, destruction of portions, or all of a space system's infrastructure cannot remove this peacetime endowment'.[101]

Enemies without space capabilities of their own can acquire them from others. America could conduct what has been called 'diplomatic space control', by encouraging states not to provide adversaries with space support during conflicts. During Operation Desert Shield, France agreed not to sell SPOT multispectral imagery to Iraq.[102]

In the event of third parties wishing to continue to sell capability to an adversary ASAT operations against the 'neutral' satellite system would become necessary.

Given the sensitivities of the situation direct physical attacks against the satellite, its ground station or associated personnel would not be the preferred first option, and information attack (IA) would be more probable. This could involve jamming, sending confusing signals to the system or introduction of a software virus. However, there are counters to such methods, and if the neutral sought to overcome them, it might leave itself open to an inevitable physical attack on the system.

For many states, space has become a crucial force multiplier. For the United States space usage may well have already gone beyond this and become a force enabler. Certainly as the military uses of space have become more important, the major powers have had to rethink their overall military doctrines to embrace the space dimension, and for some states at least the need to protect vital space assets is becoming seen as being as crucial as dominating the land, sea and air.[103]

Chapter 7

Space control

Only if the United States occupies a position of pre-eminence can we help decide whether this new ocean will be a sea of peace, or a new terrifying theatre of war.

President John F. Kennedy[1]

Space. A medium like the land, sea and air, within which military activities shall be conducted to achieve US national security objectives.[2]

Doctrinal evolution during the Cold War

Over the past 20 years, the increasing sophistication of space technology, plus the increasingly tactical, rather than only strategic use of military satellites, has led to a growing belief that space is no longer just a medium where force multiplication of terrestrial military assets takes place, but has in fact become a crucial military theatre in its own right, just as the sea and the air did in earlier historical periods.[3]

In parallel with the evolution of political attitudes towards military space, there has been a crucial evolution of USAF and Pentagon military space doctrine. Doctrine lies at the very heart of modern warfare for the advanced industrial states. 'It represents the central beliefs for waging war in order to achieve victory ... it is the building material for strategy. It is fundamental to sound judgement.'[4] According to the US Air Force, doctrine represents the central beliefs of the armed forces about the best way to wage war.[5] Doctrine is the structured thinking about military operations that guides the training, equipping and employment of military forces.

Doctrine is normally largely based on experience, and the absence of such experience contributed to the long delay in the development of a genuine military space doctrine for the armed forces of the United States, given that 'while space operations have been conducted since the late 1950s, no hostilities have ever occurred in space'.[6] In the absence of such experience, space doctrine had to be derived from theory.

The first two US doctrine documents with relevance to military space made no mention of it, other than to include it as an *environment* within the overall definition of 'aerospace'.[7] Despite the drama of the space competition with the Soviet Union during the 1960s, it was not until 1971 that the USAF first outlined the 'Role of the Air Force in Space' in its revised version of AFM 1-1, the basic doctrine of the US Air Force. American space forces were now defined as having two national responsibilities, to 'promote space as a place devoted to peaceful purposes', and to 'insure no other nation gains a strategic military advantage through exploitation of space'.[8] These roles appeared unchanged in the 1975 version of AFM 1-1.

In 1979, however, the doctrine revision significantly expanded the treatment of space operations and listed three responsibilities: to protect American use of space, to enhance the performance of land, sea and air forces, and to protect the United States from threats in and from space. The amended doctrine also identified three types of space operations, space support, force enhancement and space defence.[9]

The first *specific* space doctrine document was commissioned by the USAF Chief of Staff in 1977 and appeared in 1982. Air Force Manual (AFM) 1-6 identified three roles for space power, these being to strengthen the security of the United States, to maintain American space leadership and to maintain space as an environment where nations could enhance the security and welfare of mankind.[10] The wording reflected the continuing American division between a desire to preserve space as a peaceful sanctuary and a recognition of its potential as a theatre of military operations.

The military objectives of US space forces were described as being to maintain America's freedom to use space, to increase the readiness, effectiveness and survivability of US forces, to protect US resources from threats operating in or through space, to prevent space from being used as a sanctuary for aggressive systems by adversaries, and to exploit space to conduct operations to further military objectives.[11] The 1982 doctrine also described two existing and three potential missions. The former consisted of force enhancement and space support. The potential missions were space-based weapons for deterrence, space-to-ground weapons and space control and superiority.[12]

AFM 1-6 was updated in 1984, but there were no significant changes. The document was rescinded in 1991 in the expectation that it would shortly be superseded by an operational level doctrine for space operations. However, the envisaged document (AFM 2-25) never appeared. AFM 1-1, however, was updated once more in 1992, again with no significant changes to the existing limited space doctrine.

The United States armed forces have been slowly moving to embrace a doctrinal basis for military space operations for the past 30 years. The importance of space and its future significance have not been in doubt. In 1996 the USAF announced that 'we are now transitioning from an air force into an air and space force on an evolutionary path to a space and air force'.[13]

Soviet military space doctrine proceeded in the same direction as its American counterpart. The 1978 edition of the *Soviet Military Encyclopedia* described 'space war' as 'military operations using space and antispace resources and systems with the aim of weakening the enemy's space forces or achieving supremacy in outer space'.[14] The article goes on to describe both passive military space systems such as reconnaissance, communications, geodesy, navigation and meteorology, and active systems, such as the physical destruction or operational disabling of target satellites. The *Encyclopedia* also describes 'supremacy in space', defining it as 'a situation in which the military space systems of one side have decisive superiority over the systems of the other side. The side dominant in space is capable of performing its missions without significant enemy opposition'.[15]

The military dimension

While the Soviet space programme served political purposes, the military uses of space were fundamental throughout, and the military rationale was even more important than the political. The propaganda aspect was crucial because of the way that it related to the Soviet Union's perception of itself and its strategy to prevail in the global struggle with the capitalist West. A fundamental aspect of Soviet policy was that the USSR never regarded itself as being ultimately secure, and therefore the positive image promoted by the space programme was part of the survival strategy of the Soviet Union. But it was only part of a broader security strategy, much of which focussed on the attainment of military capability and reputation, and in the latter regard, space became increasingly important as a 'force multiplier' for Soviet military capabilities.

For most of the Cold War, Soviet leaders operated within a firmly realist ideological framework, in which international politics was seen as a critical dimension of an inevitable class-based struggle for power between the capitalist and communist states. A key component of Soviet strategic thinking was the belief that the West could only be effectively deterred from attack if it was clear that in any resulting superpower war, the Soviet Union would prevail and emerge victorious. From this perspective, while American superiority in any military-related field was inherently destabilising, Soviet advantages in contrast were a force for peace and stability. It was therefore essential that the USSR maintain at least parity, and if possible superiority, in the crucial area of military space, given its acknowledged force-multiplier effects.

The key tenets of Soviet strategic thinking in this period were outlined in Marshall Sokolovski's *Soviet Military Strategy*, originally published in 1962. In speaking of the requirement to wage a protracted nuclear war in the event of Western aggression, Sokolovski insisted that 'the essential nature of war as a continuation of politics does not change with changing technology and armament'.[16] Notwithstanding this, 20 years later Chief of the General Staff, Marshall Ogarkov, while adhering to many of Sokolovski's arguments, declared that 'technology is the foremost influence in military affairs, and military doctrine is now being driven by technology'.[17] Significantly, despite appearing only five years after the launch of Sputnik, Sokolovski's work included a section on 'The problems of using outer space for military purposes'.[18]

For Humble, the question of 'whether Soviet planners view space as a theatre of military operations, analogous to sea control, or an arena with its own intrinsic value, is debatable'.[19] The second edition of Sokolovski's book, which appeared in 1963, noted that the geographic expanse of future wars was likely to include outer space, while the 1965 edition of the *Soviet Dictionary of Basic Military Terms* defined 'space doctrine' as 'a doctrine envisaging active hostilities in space, and regarding mastery of space as an important prerequisite for victory in war'.[20] Certainly, American intelligence was convinced that Soviet military doctrine sought the same wartime goals from space as its US counterpart, namely 'to attain and maintain military superiority in outer space sufficient both to deny the use of outer space to other states and to assure maximum space-based support for Soviet offensive operations'.[21]

While the Salyut space stations made possible the international cooperation seen in the *Intercosmos* programme, they were also used for military purposes. The Salyut series in fact consisted of two separate programmes, one civilian and one military. Salyut 2, launched in 1973, was a military station, though it broke up without achieving a stable orbit. Salyuts 3 and 5 were also military stations, and flew at an altitude of 157 miles in order to make their reconnaissance missions more effective. The civilian stations, such as Salyuts 4 and 6 orbited at 217 miles up. There were also differences in the designs of the military stations compared to their civilian counterparts.[22]

Space control

The United States has not as yet had to face a military conflict with an opponent that possessed military space capabilities on a par with those of the United States, or with the ability to successfully prevent the US from exploiting its own military space assets.[23] Nonetheless, it is a reality that crucial US intelligence, surveillance, reconnaissance early warning, weapons guidance, command and control and environmental monitoring capabilities are migrating to space.[24] As they do so it clearly becomes necessary to protect them. Increasingly, control of space will become a necessary precursor to effective operations on land, sea and in the air. US reliance on space assets is becoming a 'centre of gravity' which its adversaries will target in wartime.

Adversaries with little or no ability to use space systems themselves can only benefit from attacking those of the United States.[25]

Increasingly during the 1990s, the terms 'space control' and 'space power' began to feature more prominently in key Pentagon documents. Although the use of this terminology has led to alarmist reaction from observers who see an American desire for hegemony in space, this logic does not follow from the concepts themselves.

Space control is defined by US Space Command as 'the ability to assure access to space, freedom of operations within the space medium, and an ability to deny others the use of space, if required'.[26] Space control does not require the permanent occupation or domination of space. 'We will dominate our opponent in space ... and just as our Air Force doesn't continually dominate the international skies, we haven't, and aren't going to dominate all of space'.[27] Space superiority needs to be acquired and maintained only for the duration of a specific conflict.[28] It can be seen as being analogous to control or domination of the sea lanes or air space in wartime, with an implicit assumption that these media would be returned to national and international use at the conclusion of hostilities. The United States defines space superiority as 'the degree of dominance in space of one force over another that permits the conduct of operations by the former and its related land, sea, air, space and special operations forces at a given time and place without prohibitive interference by the opposing force'.[29]

Space control can be sub-divided into surveillance, protection and negation.[30] Surveillance refers to the ability to detect, track and identify both launched and orbiting objects and determine their capacity to threaten friendly systems.

'For the first time the new National Military Strategy addresses space in terms of space power'.[31] The term 'space power' was first used as early as 1964 by Klaus Knorr, who did not however provide a definition for the term.[32] The 1998 Air Force Doctrine Document defined the concept in a fairly minimal way as 'the capability to exploit space forces to support national security strategy and achieve national security objectives'.[33] Colin Gray goes further and relates it to the capability of an adversary as well as oneself, defining it as 'the ability to use space while denying reliable use to any foe'.[34]

Oberg defines space power more widely as 'the combination of technology, demographic, economic, industrial, military, national will and other factors that contribute to the coercive and persuasive ability of a country to politically influence the actions of other states and other kinds of players, or to otherwise achieve national goals through space activity'.[35]

From this Oberg derives a list of space power elements required for a state to sustain space power. Many of these would be targets in wartime as part of a campaign to deny an adversary the ability to exploit space for military purposes. These would include space-related command and control facilities, space launch facilities and key laboratories as well as key industrial production facilities. Space hardware such as satellites and launchers would also be targets for physical destruction or degradation of capability through interference with operations.

Some targets might be attractive in any conflict, while others would only become significant in the context of a long-duration conflict. In a short-term conflict, destroying a satellite system without attacking the space power infrastructure might be sufficient. The key variable would be the adversary's capacity for reconstituting capability. In a short-term conflict attacking the ground segment of a satellite system would probably suffice to render it useless for the duration of the conflict. In a longer conflict, there might be some capacity to regenerate assets, by, for example, constructing a ground station or even replenishing a satellite. Attacking industrial targets and space-launch facilities would then become a more worthwhile strategy. A long-term conflict would require the destruction of both operational space systems and the enemy's space infrastructure.[36] 'Regardless of the type or length of an engagement, attacking the elements of space power is essential to effective counterspace operations'.[37]

AFM-1-1 describes the objective of offensive counterspace operations as being to 'seek out and neutralize or destroy enemy space forces in orbit or on the ground at a time and place of our choosing'.[38] Counterspace operations have both offensive and defensive elements. The former include operations designed to deny, degrade, disrupt, destroy or deceive an adversary's space capability.[39] The latter involve operations to preserve space capabilities, withstand enemy attack, restore space capabilities after an attack and reconstitute space forces. For proponents of weapons in space, this commitment necessarily demands that the United States acquire and deploy weapons in orbit capable of attacking and destroying an adversary's satellites.[40]

In partial recognition of this, the armed forces have begun looking at the types of technology that might be required. The US Army's Space Master Plan has identified gaps in America's space control capabilities and identified future operational requirements for space-based laser weapons, air-borne lasers and kinetic energy ASAT systems.[41]

US doctrinal evolution in the 1990s

During the 1990s, the United States armed forces increasingly looked to the significance of space as the 'new high ground'. In 1996 Air Force Chief of Staff Ronald Fogleman released a new list of 'core competencies' required of the USAF, of which the first was 'air and space superiority'.[42] This objective was repeated in AFM-1-1, the USAF's first significant effort to produce a space doctrine, which called for the Air Force to gain and maintain dominance of space.[43]

The US Air Force and Joint Staffs codified operational level space operations doctrine through Air Force Doctrine Document 2-2, *Space Operations*, and Joint Publication 3-14, *Joint Doctrine for Space Operations*. AFDD-2 deals with the command and control of space forces, both at the global and theatre levels, and the planning and implementation of space operations, again at both global and theatre levels. Especially significant in relation to the concept of space power are the sections dealing with 'the integration of civil, commercial and foreign space assets into operations'.[44] JP 3-14, which deals with joint operations doctrine, concentrates on global space forces, though there is some treatment of theatre operations.

Doctrine in the 1990s did not place significant emphasis on counter space operations. AFDD-1 Air Force Basic Doctrine stated that 'to ensure that our forces maintain the ability to operate without being seen, heard or interfered with from space, it is essential to gain and maintain space superiority'.[45] In 1991 the US had demonstrated a major asymmetric advantage with its space capabilities, but conflicts in the 1990s did not see efforts by adversaries to counter these capabilities. AFDD-1 defined counterspace operations as 'those kinetic and nonkinetic operations conducted to attain and maintain a desired degree of space superiority by the destruction, degradation or disruption of enemy space capability'.[46] It is argued that the US needs to be able to accomplish three key missions, space surveillance, space negation and space protection,[47] and that the US should proceed with the development of the technology to achieve these goals including the acquisition of ASAT systems, space mines, uplink and downlink jammers, and space decoys.

To some extent the USAF now seems to have moved to a position consonant with Lupton's 'high ground' posture.[48] Certainly the terminology is prominent in key documents, such as AFDD 2-2, which declares that, 'space-based forces hold the ultimate high ground, offering the potential for permanent presence over any part of the globe'.[49] In addition, US Space Command's *Long Range Plan* anticipated a future where 'by 2020 any ballistic or cruise missiles could be targeted, but in addition, the same space weapons could target high-value terrestrial targets'.[50]

The Space Commission Report

During the 1990s Senator Bob Smith (R-NH) was a vigorous advocate of the weaponisation of space and it was his efforts that led Congress to pass legislation included in the Defense Authorisation Bill for fiscal year 2000, which established a special Space Commission to evaluate the need for reform of US military space organisation and capabilities.[51]

Donald Rumsfeld chaired the commission until he was nominated by President Bush to serve as Secretary of Defense as the commission was finalising its Report. As Secretary of Defense, Rumsfeld was able to ensure that many of the commission's recommendations were implemented. As a result a single military service, the USAF, has become the DoD's executive agent for space, with the Under Secretary for the Air Force assuming direct responsibility for all national security space, including the National Reconnaissance Office.[52] In March 2001, a Space Policy Coordinating Committee was established under the National Security Council.

The Commission reaffirmed the traditional American commitment to the peaceful uses of space declaring its 'conviction that the US has an urgent interest in promoting and protecting the peaceful use of space'.[53] Nevertheless, it also called for the development of physically destructive anti-satellite capabilities and the development of 'live firing ranges' in space to test these systems on a regular basis.[54]

The Report recommends a general ignoring of alleged legal impediments to the use of weapons in space. It does this by asserting that the US and most other nations interpret 'peaceful' to mean 'non-aggressive', and that non-aggressive incorporates the legitimate right of self-defence, including 'anticipatory' self-defence under the UN Charter and Article III of the Outer Space Treaty. In addition it notes that 'there is no blanket prohibition in international law on placing or using weapons in space, applying force from space to earth or conducting military operations in and through space'.[55]

Current space doctrine

The merger of USSPACECOM and USSTRATCOM to form the new USSTRATCOM meant a requirement for updating of the space doctrine, a process encouraged also by experience derived from operations in Afghanistan and Iraq. The merger meant that the new joint space support teams would integrate all STRATCOM missions, including space, global strike, global ISR, information operations and missile defence.[56] USSTRATCOM carries out its functions through four primary missions, space support, force enhancement, force application and space control. Space support refers to the operations needed to enable space capability to be exercised, for example space launch and satellite operations. Force enhancement refers to the force multiplier effects to terrestrial forces that have become familiar over the past 30 years, such as intelligence gathering, early warning, communications, navigation and weather forecasting. In these functions the armed forces are supplemented by capabilities from civil, commercial and national space systems. Force application involves applying force either from or through space. Space forces can target land, sea and air forces. They can do this either by acting as the 'gunsights' for terrestrial weapon systems, or by directly attacking terrestrial forces with space to ground weapons.

To an overwhelming extent, the purpose of acquiring space control in wartime is in order to achieve 'information dominance'.[57] By maximising one's own ability to acquire and correlate crucial military information, while denying similar high quality to the enemy, the latter's task is made overwhelmingly difficult. In order to achieve such a position in wartime, and prepare for it in peacetime, US analysts recognise that 'the final ingredient of a true space doctrine is an explicit statement by the national leadership that space is no longer a sanctuary but rather the high ground of a global *infonet* which can be used for civil or military purposes'.[58]

The operations in Afghanistan and Iraq showed that the existing space operations doctrine provided inadequate detail regarding the coordination and integration of space forces supporting theatre operations. In addition, Iraq's attempts to jam US global positioning system signals in 2003 showed that US adversaries had understood the importance of US military space capabilities and were beginning to develop capabilities to counter and disrupt them.[59] Even though the Iraqi efforts were not successful and were defeated by GPS guided munitions, the experience reinforced the requirement to develop a doctrine relating to counter-space operations.

For advocates of space power, the operations from 1990 onwards validated its potential. Flavell, for example, argues that the GPS satellite system enabled 'dumb' bombs to become accurate all-weather weapons. 'Operation Allied Force highlighted the synergy of these new "space aided" weapons; the enemy could no longer rely on weather as a "sanctuary"'.[60] During Operation Iraqi Freedom 'US forces conducted pre-emptive strikes on Iraqi leadership based on real-time satellite feeds to the cockpit'.[61] 'Due in large part to space systems, US military forces know more about their adversaries, see the battlefield more clearly, and can strike more quickly and precisely than any other military in history. Space systems are inextricably woven into the fabric of America's national security'.[62]

In August 2004 the United States Air Force published AFDD 2-2.1, *Counterspace Operations*, the first doctrinal document on this critical subject. USAF Chief of Staff General Jumper noted that USAF doctrine was 'evolving to reflect technical and operational innovations'.[63] The rationale for the doctrine is succinctly presented at the outset, as following from the same military logic as the requirement for air superiority, to gain control of the skies at the outset of a campaign and deny them to the enemy.

AFDD 2-2.1 declares firmly that 'space superiority provides freedom *to* attack as well as freedom *from* attack'.[64] Counterspace embraces both offensive and defensive operations, both of which are dependent upon effective space situation awareness (SSA). Defensive counterspace operations preserve the US ability to exploit space for military purposes and include passive satellite defences, such as the use of camouflage, concealment, deception, dispersal and the hardening of systems. Offensive counterspace operations are those designed to deny an adversary use of space to support their military operations. The methods employed to do this may be permanent or reversible and embrace the 'five Ds' – deception, disruption, denial, degradation and destruction. The more dependent an adversary is on space capabilities, the more vulnerable it is to counterspace operations. Offensive counterspace operations may target the ground or space segment, or the links between them. Ground segment may include both ground stations and launch facilities. Methods used may range from laser weapons to special operations forces. Execution of effects is in terms of the Space Tasking Cycle, which turns the force commanders' priorities and intentions into a coherent plan for the use of space forces. Space Tasking Orders, usually disseminated 6 hours prior to implementation, tasks Space Command forces for the next 24-hour period.

US counterspace operations, like all other US military operations, reflects an effects-based methodology, to allow the choice of the tactics most appropriate to achieving the objectives. Among other things this requires careful planning to ensure that objectives at every level, tactical, operational and strategic are taken fully into consideration when planning counterspace operations. AFDD 2-2.1 makes the important points that neither adversary use of space nor counterspace operations necessarily require that adversaries be space-faring nations themselves. US space capabilities can be attacked at the ground segment as well as the data links, and the space segment can be attacked by weapons fired from the earth. Similarly, adversaries can purchase space services and products such as imagery and communications.[65]

A catch-22 of the pursuit of space control is that it is to some extent an all-or-nothing strategy. If an adversary is a non-space using opponent, or if its space assets are quickly lost through offensive counterspace operations, then it would have nothing to lose by conducting major ASAT attacks, at which point defensive counterspace capabilities would be at an absolute premium to its opponent. Given that such attacks might be electronic rather than physical, and even when physical, aimed at the ground segment rather than the satellites themselves, they might well prove extremely difficult to defend against. Such asymmetric space conflict would be particularly problematic against a nuclear armed opponent, since it would have the option of launching a nuclear weapon to detonate in the upper atmosphere. The consequent electro-magnetic pulse and ionisation effects would be likely to disable satellites over a wide area.[66]

As with AFDD 2, the new Counterspace doctrine divides space forces into three categories, global, deployable and theatre organic. Global space forces are those that support national objectives and multiple theatres, such as the GPS radio-navigation system. Deployable assets are those which must be moved in theatre to support operations, such as the JTAGS system that provides commanders with downlinked, in-theatre early-warning of ballistic missile launches. Organic space forces are those normally deployed in theatre, such as Eagle Vision that provides real-time acquisition and processing of commercial satellite imagery.[67]

Non-space ASAT options

Proponents of the acquisition of ASAT capabilities by the US argue that ASAT weapons are essential if the US is to dominate the space environment, because they are essential to the protection and negation roles.[68]

However, it is important to remember that ASAT does not need to be space-based, or necessarily to involve the physical destruction of the satellite. Jamming, spoofing and control seizure can be done from the ground, and terrestrial elements of the satellite system, such as ground control, can be attacked by conventional terrestrial methods. DeBlois *et al.*, after a detailed survey of possible technologies, argue that 'space weapons are generally not good at protecting satellites' capabilities'.[69]

The objective of US space control is essentially information dominance in wartime. The Commander of US Space Command after the 1991 Iraq War declared that space forces reduce the problems caused by Clausewitz's 'fog of war' and make the battlespace more transparent. 'With space forces, we can rapidly observe, hear, understand, and exploit a battlespace environment anywhere in the world, even in remote locations, with little or no local support infrastructure'.[70] In order to achieve this it is the information flow that is critical and not necessarily the information systems themselves. Current US space control thinking tends to focus on physical assets, rather than on capabilities.[71] Information dominance rather than asset destruction is the requirement. Space control is about dominating the space lines of communication, and for this the requirement is simply to impact effectively upon one of the segments of the space system or the links between them.

In 1999 Deputy Secretary of Defense John Hamre testified before Congress that DoD views on space control emphasised the temporary denial of space to an enemy, rather than the destruction of space systems.[72] Since space is a global commons, return of full access to space for all nations as soon as possible must be part of the 'exit strategy' for space operations in wartime.[73] This requires the US to possess the full spectrum of military options for counterspace operations (lethal to non-lethal) and a doctrine that produces desired effects with minimum impact on the commons. The 2001 Space Commission report re-emphasised this, noting that, while the US reserves the right to destroy either ground sites or satellites if necessary, the preferred approach is to use methods that are 'temporary and reversible in their nature'.[74]

Offensive counterspace, aimed at denying the enemy the use of space in wartime, can be carried out in three ways. First, targeting the enemy's terrestrial space segment, their launch infrastructure, satellite command and control systems and satellite communication nodes.[75] These capabilities are already possessed by the United States. The second approach would be to target the communications segment between the satellite and its associated ground equipment. The United States currently has the capability to successfully jam the ground segment, but has no capacity to interfere with the space segment.[76] The third approach would be to launch a direct physical attack against the satellite itself.[77] The United States does not currently have the capacity to carry out such attacks, except by using nuclear warheads, which is against a number of treaties to which the US is a party and which would create damaging debris and EMP effects in the satellite orbit and upper atmosphere, or by using residual ASAT systems, such as the Space Shuttle, which are not optimised for such a role.

AFDD-1 states that the United States must achieve and maintain space superiority, but at no point does it suggest that space weapons are required to do so.[78] There are a variety of ways to achieve the same end, including 'implementing an international agreement to shut off a satellite's downlink, terminating imagery sales, destroying ground sites, destroying or disrupting system software programs, spoofing or jamming link signals, damaging or disrupting satellite subsystems, and disabling or destroying the satellite'.[79] The use of ASAT weapons is the least attractive option. 'Anti-satellite weapons may have been the only method to achieve space control in the early decades of space exploitation, but they are not as viable in today's information dominated society'.[80]

The political context

The United States continues to maintain a dualistic posture on space policy. US Undersecretary of the Air Force, Peter Teets, argued that 'Having come to rely on the unhindered use of space, Americans will demand no less in the future. This reliance demands the continuance of robust capabilities for assured launch and space control. Although the United States supports the peaceful use of space by all countries, prudence demands that we ensure the use of space for us, our allies and coalition partners, while denying that use to adversaries'.[81]

A point in regard to the whole weaponisation debate is that it makes a curious geographical distinction. In a real sense, space is both militarised and weaponised. There are no space weapons targeted against inanimate pieces of orbiting machinery, but it is not true that space does not have weapons that directly impact targets. As the Commander of Air Force Space Command put it in February 2004, 'the Taliban and Iraqi Republican Guard forces on the receiving end of satellite guided weapons are likely to have a different impression. Many of today's weapons employed by the US Army, Navy and Air Force are targeted by overhead space systems, commanded using space connectivity, and guided by precision space based navigation systems'.[82]

DeBlois argues that while proponents talk as if America could achieve total space dominance for a variety of purposes, the reality is a reactive environment. 'Principal powers will simply not allow a space hegemon to emerge, and lesser powers may concede hegemony but will continue to seek asymmetric counters'.[83] ASAT weapon attacks might indeed prove to be counterproductive to US interests, leading to the escalation of the conflict, damage to crucial American space assets, and deterioration of US relations with its allies.

While there may be clear military rationales in favour of the weaponisation of space by the United States, it is a decision that would have considerable political implications. It is also true that to date there have always existed powerful cultural and political domestic obstacles in the United States to such a development. Even at the outset of the space age leading US politicians speculated on the idea of space as a force for peace rather than a theatre of war. House Majority Leader McCormack suggested in 1958 that the exploration of space had the potential to encourage a revived understanding 'of the common links that bind the members of the human race together and the development of a strengthened sense of community of interest which quite transcends national boundaries'.[84] President Kennedy similarly suggested that it was 'an area in which the stale and sterile dogmas of the Cold War could be literally left a quarter of a million miles behind'.[85]

US National Space Policy states that the United States is committed to the exploration and use of outer space 'by all nations for peaceful purposes and for the benefit of all humanity'.[86] US national space policy does allow for the use of space for the purpose of national defence and security, but nevertheless, the weaponisation of space would seem to run counter to a very long-standing national policy.

Similarly, the US National Security Strategy declares that uninhibited access to space and use of space are essential to American security. Space policy objectives include protecting US space assets, 'preventing the spread of weapons of mass destruction to space, and enhancing global partnerships with other space-faring nations across the spectrum of economic, political and security issues'.[87]

It is also notable that the US armed forces are aware of the need to respect the concept of space as a 'global commons', so that if 'the United States impedes on the commons, establishing superiority for the duration of a conflict, part of the exit strategy for that conflict must be the return of space to a commons allowing all nations full access'.[88] Current US military space doctrine is careful to emphasise the political implications of military operations in space and the need to be sensitive to legal issues. USDD 2-1.1, *Counterspace Operations*, insists that 'in all cases, a judge advocate should be involved when considering specific counterspace operations to ensure compliance with domestic and international law and applicable rules of engagement'. [89]

Nevertheless, the implications of anti-satellite operations in time of war have profound implications for the routine conduct of international relations by those states not party to the conflict. During the 2003 US–Iraq war, Iraqi forces attempted to jam US satellites but were not successful. Given the likelihood that future conflicts are increasingly likely to involve states with satellite-based force multiplication, ASAT operations will be routine and since most states use systems for which they are neither the originator nor the only user, negating a satellite to block one state's access will necessarily impact on the capabilities of other states not party to the conflict. Under international law, the United States 'cannot limit access to space by any nation, much as it cannot keep ships off the seas or planes out of the air. In times of war such limitations succumb to national security imperatives'.[90]

Space begins where laws change. Where international law replaces domestic laws of national sovereignty and where the laws of orbital mechanics take over from the laws of aerodynamics. International law would clearly mitigate against a move to weaponisation. The 1967 United Nations Outer Space Treaty states that 'outer space, including the Moon and other celestial bodies, is not subject to national appropriation by claim of sovereignty, by means of use or occupation or by any other means'.

The lack of response by the international community to the growing threat to space as a sanctuary suggests that the United States would not significantly alienate itself from the international community if it crossed the threshold to the weaponisation of space. There has been no reaction to the doctrinal evolution over the past decade, and the ultimately submissive reaction to the US withdrawal from the ABM treaty is suggestive.

Military commentators question the use of diplomacy as a method for protecting America's satellites on a number of grounds. One is that arms control agreements regulate the peacetime environment, but would not be honoured in wartime.[91] Five years before the first flight by the Wright brothers, the Hague Peace Conference banned the use of aircraft, existing or projected, for combatant use in war. Their use was restricted to passive employment such as reconnaissance. But the air warfare of World War One invalidated these restraints. It is suggested that 'similar standards for the strictly peaceful use of space (reflected in many international treaties) are likewise facing the inevitability of war in space.[92]

A second argument is that the United States currently enjoys a massive military space superiority over other states, and that it should not therefore support treaties that would erode or remove that advantage, and create a level playing field for its future adversaries. 'The US would never negotiate away air or sea superiority. It makes no more sense to allow an enemy unfettered access to space'.[93] Certainly during the Cold War, during the periods when the US was a strong proponent of arms control as a way of contributing to US national security, it was extremely clear that naval arms control was not on the agenda, for the same reasons that the US might resist space arms control in the current period.

Certainly the United States has shown a consistent reluctance to support efforts to develop a more constraining arms control regime for space, and has argued at the UN Conference on Disarmament that it sees the current international space regime as entirely satisfactory and in no need of renegotiation.[94]

The United States could, at some cost, place weapons in space. That it has not yet done so is because such a step would be in conflict with long-established national space policy. The existing US national space policy is the main barrier to the weaponisation of space, and is 'a remnant of space policies developed during the Cold War'.[95] The non-weaponisation of space is thus primarily due to an American self-denying ordinance, not primarily to commitments imposed by international law, and the international community currently lacks the capacity to influence the American decision-making process in this regard.. A US decision to cross the threshold is likely to be contingent on the actions of other states, most notably China.

Chapter 8

Space, justice and international development

It is easy to think of the politics of space in realist terms, as being simply the result of the interaction of powerful states, primarily concerned with their military security. In reality, however, the issue of international development is also at the heart of debates about space policy. By the very nature of the subject, debates about space policy encourage a holistic approach, in which the earth and humanity as a whole are central to understanding. The same is true of debates about development in international relations,[1] and the two have been linked throughout the space age.

The North–South system of international political and economic relations is characterised by dependency and inequality. The states of the developed world argue that this system, while not perfect, is legitimate, because it produces benefits that apply to the system as a whole. The states of the developing world, in contrast, feel that the current international system is essentially illegitimate, because it was structured by imperialism and colonialism and works to hinder their development and ensure that the benefits of the system flow overwhelmingly towards the rich states of the developed North. Not surprisingly, therefore, the developed states consistently work to maintain the *status quo*, while the South seeks to replace it. This basic conflict of interest forms the fundamental dynamic of North–South international relations.[2]

In the 1960s the newly independent states of the developing world began to work together in bodies like the United Nations to press the developed states to make changes in the management of the international economy and in particular the operation of the international trading system. Having gained political independence from their former colonial masters, they had discovered that they were far from being equal members of the international community. The structural inequalities of the international system operated in such a way as to make it extremely difficult for them to develop rapidly in order to attain the standards of living existing in the developed states. They therefore called for the creation of a 'New International Economic Order' in which they would gain a greater share of the world's wealth, and a greater role in global policy making. Of even greater relevance in terms of the importance of telecommunications and remote sensing satellite technology was the related call for a 'New World Information and Communication Order', which was the subject of intense debate within UNESCO. In 1991, when a group of developing states submitted proposals at the United Nations for a reordering of the system of international space cooperation, their arguments drew heavily on the language of the NIEO and NWICO.[3]

Space technology has become increasingly important to many developing and newly industrialising states, and is seen by some as a way to bypass intermediate stages of development, and at the same time become more independent of the developed industrialised states. The economic benefits of space technology have a dramatic 'multiplier effect' on a state's gross national product.[4] However, developing countries face four major difficulties in joining the 'space club'. Compared to the developed countries, they lack capital, have far fewer technically skilled personnel, have a much weaker scientific support base, and may not always have the stable political system and policymaking apparatus that is required to sustain the long-term political and financial support of a successful space programme. These weaknesses also make it more difficult to establish stable multilateral associations for the development of space technology and applications. One solution is to import resources from outside, but this may be difficult for various financial, technical and political reasons, and even if possible, undermines the rationale of escaping from dependence on the developed countries. In practice, access to space is currently controlled by a small minority of the world's 200 states,[5] though the number of launch sites in the developing world is increasing.

International law clearly states that outer space cannot be legally appropriated by any one actor because it is *res communis*, that is, owned communally by mankind. Furthermore, the Outer Space Treaty of 1967, in its first article, states definitively that outer space is the 'province of all mankind' and in article 2, that its exploration and use is to be carried out for 'the benefit of and in the interests of all countries'. From this and other practices of both international convention and law 'it is assumed that all states are entitled to participate in decisions regarding its use and to share in the economic payoffs from exploiting it'.[6]

In practice, however, access to and benefits from space are not shared in the manner that this suggests. In particular, from the 1970s these assumptions began to be challenged by the increasingly assertive developing nations who were demanding a 'new world economic order' in which they could benefit from a more equitable distribution of the world's resources and preferential trade practices. While international law proclaimed equal access to space, the reality, according to the developing nations, was that the developed world would easily dominate the exploitation of space, due to their huge technological and economic superiority. This disadvantage was compounded by the practices of the international community, which gave preferential treatment to the developed states in the implementation of international legal agreements regulating space resources. The net effect of this discrimination, it was argued, was to deny the poorer countries of the world access to space, and thereby prevent them from exploiting one possible path to accelerated development. It was an issue of such importance that unless addressed 'conflict in the international system between the developed and underdeveloped nations will steadily and substantially increase'.[7]

Space technology offers enormous potential benefits to developing nations. Examples of such uses include improvements in communications, particularly for large countries with difficult terrain, in education, where satellite technology can dramatically accelerate a government's ability to provide education to large and disparate populations, particularly in rural areas, in meteorology, where vital rainfall levels could be monitored and severe weather conditions predicted and tracked, for dealing with the effects of natural disasters and for identifying and exploiting agricultural, geological and hydrological resources.[8] Satellite remote sensing, in particular, has enormous existing and potential value for developing nations. For example, 'because the agricultural production and distribution system reacts slowly to emergency conditions, advance warning of drastic changes in predicted crop production is necessary to prevent either famine or oversupply. For this purpose the synoptic view of satellites is especially advantageous'.[9] Satellite remote sensing can benefit developing countries not only in terms of economic and particularly agricultural information, but also with regards to education provision, health services, conflict resolution, environmental protection and crime prevention.[10]

Satellite remote sensing is the use of space-based sensors normally operating at wavelengths from the visible to the infrared, in order to collect data about the Earth's atmosphere, oceans, land and ice surface. The first photographs taken from space and successfully retrieved were those of the US Explorer 6 satellite launched in August 1959. This was the forerunner for spy satellites, meteorological satellites and Earth-resources satellites. The first formal remote-sensing experiment was conducted on the Gemini 4 manned mission in June 1965. It was a successful geological survey, which encouraged further resource surveys on subsequent Gemini and Apollo flights. These missions encouraged the US Geological Survey to lobby successfully for a dedicated Earth resources observation satellite programme, which was subsequently inaugurated with the launch of Earth Resources Technology Satellite 1 (ERTS-1, subsequently renamed LANDSAT-1), launched in July 1972. Subsequent LANDSAT satellites carried a wide array of different types of sensors and the programme was extremely successful, with LANDSAT data being purchased by a number of countries, including developing states such as India. India went on to develop its own remote-sensing satellite, Insat, which was launched in 1982 on a Soviet launcher. The USSR, China, Japan, France and the European Space Agency all orbited remote sensing satellites in the 1980s and 1990s.

In the 1980s, for example, Thailand initiated new controls on logging and a major replanting programme after data provided by the US LANDSAT satellite indicated that there would be no rainforest left by the early twenty-first century if the current rate of deforestation monitored from orbit had continued. Sensors in space can measure soil temperature and moisture content, providing data which allows planting to take place at the optimum time. Once planted, fields can be monitored from space for potential threats such as drought, flood and disease. LANDSAT imagery was used to identify host plants of the parasitic Mediterranean fruit fly during 1979–80, which later led to their controlled destruction.

The ability of satellites to measure the snowline on mountains has become a crucial tool of flood control, since it is linked to water run-off volumes. Satellites are also now routinely used to detect underground water sources from orbit, which is central for planning irrigation systems and for managing ecologies in areas where rainfall is limited and drought conditions common.

Other uses include the measurement of erosion, the planning of roads (this was done in Upper Volta as early as 1980), urban development planning, cartography, forestry management, pest control and wild land evaluation. The value and utility of space derived monitoring data is almost limitless, and has transformed the effectiveness of countless applications.

Satellites have applications that go well beyond the uses a particular state might have for them. Satellite sensors are capable of mapping areas and detecting patterns that go beyond national concerns. Two examples of this are atmospheric ozone measurement and observation of the continent of Antarctica, both of which were central to the identification of the 'greenhouse effect' and global warming phenomena.

While space applications have enormous potential for developing nations, it is not necessary that they have fully-fledged space programmes in order to benefit, since they could simply access data from other national or commercial satellite systems that are most appropriate for their concerns.[11] The application of remote sensing and communications technology have been heralded both as an enormous benefit for humanity and as a method for maintaining the international *status quo*. Central to this debate is the concern shared by many developing countries that remote sensing and other space technology are merely tools of global neo-imperialism.

President Nixon told the UN General Assembly in 1969 that the United States had 'decided to take actions with regard to Earth resource satellites. The purpose of those actions is that this programme would be dedicated to produce information not only for the United States, but also for the world community'.[12] From the outset it was clear that remote sensing technology had as much if not more to offer the developing countries than it did to the developed world. By 1980, 120 countries had purchased LANDSAT data from the United States, of which nearly two-thirds were developing states. Throughout the developing world, government bureaucracies have been established to process satellite data or to conduct and control remote sensing operations. The fear of exploitation by the developed states emerged because information is power and remote sensing offered the power to develop resources. Many developing countries feared that since they would not be in control of the dissemination of sensed data their dependency on the developed world would simply be reinforced. In particular, they feared that they would be placed at a disadvantage in negotiations with multinational corporations, who would have access to satellite data which would put them in a superior bargaining position when negotiating for rights to exploit resources.

This assumption has become weaker over time. Some countries, such as Brazil, India and Indonesia, have acquired their own remote-sensing capabilities as well as purchasing data from a variety of different satellite data providers and their growing experience in this field has seen the information disadvantage disappear. Over time a much larger number of countries have constructed ground receiving stations for satellite data, often with assistance from external sources such as the US Agency for International Development. Other countries have acquired equipment and trained their staff to interpret remotely sensed data acquired from external providers. In addition, a number of important global space data networks provide their information for free. Examples of this are the World Meteorological Organization, to which national weather satellites transmit data, and the Global Terrestrial Observing System developed by the WMO and a number of UN bodies. GTOS links existing monitoring sources and databases to provide freely data on terrestrial ecosystems and socio-economic forces, and gives priority in its organisation to the needs of developing countries.[13]

Some developing countries (as well as certain developed states) also had concerns about remote sensing as an infringement of national sovereignty. Developing states feared that remote sensing would allow their natural resources to be located and exploited by outside powers, possibly even in a clandestine manner. The view was that to some extent non-cooperative remote sensing may be seen as a form of spying, and in an era of economic competition and environmental security concerns, 'spying' need not be restricted to observation of another country's military capabilities.

A related issue that raised wider concern was the question of dissemination of information regarding one country to a third party without the consent of the latter. France and the Soviet Union were among the countries who exhibited worries on this question. At issue was the question of whether a country is entitled to exclusive jurisdiction not only over its own resources, but over information about those resources. For many countries domestic laws regard such information as privileged government data. Other states believe that imagery portraying their natural resources and their physical geography is a strategic interest.

The impact of satellite technology on states is not limited to military and resource security. States have also shown concern about unwanted political, cultural, commercial or religious messages being beamed into a country without the permission of the local government. This is not simply a fear of totalitarian regimes, even open democratic societies may fear the submerging of their cultures.

In contrast to the natural concerns such monitoring raises for certain states is the history of remote sensing itself. Remote sensing was made possible by the initial development of reconnaissance satellites for military purposes. These 'national technical means' of verification, as the SALT I strategic nuclear weapons treaties described them, made it possible to conclude safely a range of arms control and confidence-building agreements from the early 1970s onwards that played a significant part in both stabilising the Cold War and subsequently bringing it to an end. Thus while remote sensing can threaten states by revealing their military capabilities and deployments, it can also help them become more secure, by providing mutual reassurance about the absence of feared threats. This proved of historic importance in Europe in the late 1980s and early 1990s and can be applied to other unstable political relationships around the world.

Having the equal right to explore and exploit space does not mean the same thing as having equitable access to it and the benefits that can be derived from that access. Equitable does not mean equal, rather in legal terms it means justice and fairness in relation to the facts and circumstances of a particular case. Equity depends on circumstances and international agreements do not include clear definitions of equity. The stress on equity by the developing nations is related to their desire to have a greater share of global material resources.[14]

An example of this is the 1979 Moon Treaty, which calls for the establishment of an international regime that would, among other things, provide an 'equitable sharing by all states parties in the benefits derived from (lunar) resources, whereby the interests and needs of the developing countries, as well as the efforts of those countries which have contributed either directly or indirectly to the exploration of the Moon shall be given equal consideration'. However, even in this treaty, where the developing nations are clearly being included as equal partners in the agreement, they are only granted an equitable, that is non-equal share of the benefits. More disturbingly, the Moon Treaty, which came into force in July 1984, was not ratified by the United States or the Soviet Union. The regime that was intended to allow the sharing of the Moon's resources would clearly remain impotent as long as the means to carry out such exploitation were in the hands of the major powers who refused to be party to the agreement.

The entire history of the space age has been characterised by similar inequalities and discriminations. The developing world has the potential to benefit enormously from the exploitation of space technology, particularly in terms of communications and earth resources satellites, but it is precisely this area where discrimination by the developed space-faring nations remains strongest.

Since the beginning of the space age the United Nations has exercised a special role as a focal point for international cooperation on the peaceful uses of space. It has also been the forum wherein was developed an entirely new body of international law governing the exploration and use of the 'final frontier'. The UN was quick to recognise the potential importance of space exploration to terrestrial international politics. As early as 1956 the United States proposed to the UN General Assembly that Earth satellites and space platforms should be subject to international inspection and participation to ensure that such developments in outer space should be devoted exclusively to peaceful and scientific purposes.

The launching of Sputnik in 1957 led the United Nations to form the Committee on the Peaceful Uses of Outer Space. The Committee was made a permanent UN body in 1959, and has subsequently become the key UN instrument for the development of the international law of space.[15] The new committee was requested to 'review, as appropriate, the area of international cooperation and study practical and feasible means for giving effect to programmes in the peaceful uses of outer space which could appropriately be undertaken under United Nations auspices'.

Developing countries see the United Nations as 'their preferred agent for deliberation and guidance for space affairs as well as a forum in which to express their political views. Specifically, the UN Special Political Committee, under whose administrative management the Committee on the Peaceful Uses of Outer Space (COPUOS) functions, is the focus of their hopes and aspirations, fears and concerns, with respect to space. It provides the major forum for space-related issues – new regulations, proposed restrictive regimes, and challenges to Western world policies, politics and business practices. If COPUOS, which operates by means of consensus, fails to reach agreement on a given course of action, the Special Political Committee, which is dominated by the developing countries, may refer matters to the General Assembly for action.

From the dawn of the space age in the late 1950s, international law and political rhetoric has subscribed to the view that the benefits derived from the use and exploration of outer space should be shared among all nations, including the developing countries. Even the landmark International Geophysical Year in 1957–8 seemed to suggest this. Half the participating nations were from the developing world. However, many of them were included without having any active involvement. Some were included only on the basis that they were allowing observation teams and technology from developing states to be established on their territories. The developing states lacked the financial, technological and human resources to participate in a genuine manner, and the political and socioeconomic tensions remained intact and served to hinder any genuinely cooperative or collaborative efforts.[16]

On 20 December 1961, the UN General Assembly adopted a resolution initiated by COPUOS which declared that 'the exploration and use of outer space should only be for the betterment of mankind and to the benefit of States irrespective of the stage of their economic and scientific development'. The United Nations Space Applications Programme was established to provide technical assistance to developing countries, and several working groups were formed, one of which was to study the issues relating to remote sensing.

The UN took the lead in establishing certain guidelines by which states were expected to abide in their activities in space. Two key elements of this regime were that international law, including the UN Charter, were held to apply to outer space and celestial bodies. A second central tenet was that outer space and celestial bodies were free for the exploration and use by any and all states in conformity with international law, and that they are not subject to national appropriation.

During the course of the 1960s COPUOS concentrated on the elaboration of basic legal principles governing the activities of and cooperation between states in the exploration and use of space. It also played a central role in the development of specific agreements such as those dealing with state liabilities for accidents and problems caused by spacecraft returning to earth, and on the provision of assistance to, and safe return of, astronauts and spacecraft. It was strongly supportive of the separate agreement under which the USA, UK and USSR agreed to a treaty banning the conducting of nuclear tests in the atmosphere, under water and in outer space. The most significant achievement from the UN process, however, was the conclusion of the Outer Space Treaty in 1967.[17]

Later meetings of COPUOS indicated the dissatisfaction of the developing world with the Outer Space Treaty. Article one was seen as imposing a moral obligation on signatories, and developing states sought to make the obligation for space cooperation a more formal commitment under international law. Limited progress in this regard was made in the 1980s. After discussions lasting nearly a decade, the UN General Assembly agreed a set of principles. These largely reaffirmed the wording of article one, but did provide that nations should have access to data received on their territory and provisions were made for the processing and analysis of data for natural resources management and environmental monitoring. The special needs of the developing nations were, to a limited extent recognised.[18]

It seems paradoxical that empty space could be considered a 'resource', and that any part of such a vast near-vacuum might constitute a 'scarce' resource. The explanation for this is that in order to perform their functions, satellites must travel in particular orbits, and some orbits are more useful for certain purposes than others. The most useful orbit for communications satellites is the geosynchronous Earth-orbit. Similarly, satellites use radio waves to communicate, acting as relaying stations to allow signals to be beamed anywhere on the Earth's surface. Some frequencies are more efficient and easy to use for certain purposes than are others.

There are limits to the numbers of users that can be accommodated by radio waves and by the GEO, since in both cases overuse and misuse can result in users' signals interfering with one another. Unlike other 'natural resources' such as oil reserves, fisheries or fertile soil, it is impossible to deplete or permanently damage either radio waves or orbital space. Once the misuse of these resources is discontinued, they resume their original condition and are of undiminished value to future users.

The most heavily utilised and important space communications are the fixed satellite systems (FSS) and the use of the geosynchronous earth orbit (GEO) for communications is the most commercialised aspect of space activity. The GEO is a tunnel-like ring of space approximately 36,000 kilometres above the Earth's equator. It has a fundamental feature that makes it a highly valuable and desirable resource. Satellites placed in orbit in the GEO take exactly 24 hours to complete one orbit. Since the Earth completes one rotation in the same period, the relative position of the satellite above the Earth's surface remains unchanged, making it appear to be motionless, though it is simply matching its speed with the planet. Ground-based receivers can thus receive signals from the satellite without having to track the satellites' motion across the sky, making it ideal for communications systems such as television and radio, as well as meteorology and navigation. In addition, because about 40 per cent of the Earth's surface can be covered from a single location in GEO, only three satellites are required for global coverage. Other orbits require much larger numbers of satellites to achieve the same coverage. Because of this 'nearly every country in the world, either developing or advanced, has an absolute interest as to the proper utilisation of the orbit'.[19] By both tradition and international treaty, the GEO is considered to be part of the 'global commons', the common heritage of mankind.

The organisation responsible for regulating this resource is the International Telecommunications Union (ITU), the oldest universal membership organisation in international relations.[20] The ITU allocates frequencies and sets technical standards for operating procedures and equipment.[21] The manner in which it does this is highly complicated, and the very complexity of these arrangements has 'disguised the fundamental issues involved, and provided cover from the public glare of political discourse'.[22]

The GEO is thus a much sought-after orbital position for satellites. Since satellites interfere with each other's electromagnetic signals if they are too close to each other, there are limits to the total number of satellites that can operate in the GEO at any one time, though the precise number is affected by technological developments. This means that there is a 'congestion' issue relating to the GEO, particularly with satellites operating at certain frequency bands.[23] This congestion creates the need for a strict international regulatory regime in relation to the GEO, but developing countries argue that the existing regime is discriminatory and unacceptable, because it allows the more technologically advanced states effectively to occupy permanently and dominate what the International Telecommunications Union describes as a 'limited natural resource' to the exclusion of other states. From the perspective of the developing states the GEO orbital slots were a natural resource that was 'being used up before their very eyes'.[24] According to the 1973 ITU Convention, both the radiofrequency spectrum and the GEO are limited natural resources and must be used in a manner that is efficient and economic so that countries may have equitable access to both.[25] However Article 33 of the ITU Convention also states that equitable access must be in relation not only to each country's needs, but 'to the technical facilities at their disposal'. The GEO therefore presents an exquisite international political dilemma, 'how can equitable and beneficial access to the geosynchronous orbital arc be assured to all countries, be they less developed, industrialising or highly developed information societies and, in particular, should such access be based on opportunity, resources or need?'[26]

In 1976 a group of eight equatorial states (Brazil, Columbia, Congo, Ecuador, Indonesia, Kenya, Uganda and Zaire), issued the Declaration of Bogota, in which they asserted a territorial claim over the portion of the GEO immediately above their respective territories.[27] They claimed that the GEO was an established physical reality and a limited resource over which they were entitled to exercise sovereignty. Since there was no entirely satisfactory definition of where the Earth's atmosphere ended and 'outer space' began, it was not necessarily true that the GEO was in outer space and that therefore national domestic law could still apply to it. This being the case, the terms of the Outer Space Treaty forbidding national appropriation did not apply to the GEO. In particular, they rejected the idea that there could be a convention allowing a right of succession in regard to satellites.

The invocation of the sovereignty argument by the Bogota states was ironic to some extent, because the whole thrust of the critique of the developed world has essentially been based on an ethical argument, that all human beings are 'moral equals, regardless of the accident of nationality'.[28]

The terms of the Declaration were a clear violation of Articles I and II of the Outer Space Treaty, and indeed if accepted would have destroyed the crucial international principle regarding non-appropriation of territory beyond the Earth's atmosphere conventionally accepted as being at 100 kilometres above the Earth's surface.[29] Nevertheless, the underlying motivations behind the claims of the equatorial states were significant. Nor could they be dismissed without thought, since they occurred in the same period that a number of states were unilaterally and illegally extending their maritime jurisdiction from 12 to 200 miles, yet international law subsequently changed to recognise the validity of most of these declarations. The Bogota Declaration raised serious issues therefore.

There was clear validity in some of the arguments, for example, that proposed ITU solutions to key difficulties were unfair and impracticable for the developing countries, and that both the GEO and the radiofrequency spectrum had been administered in a way that did not in practice allow 'equitable' access for the developing countries. Outer Space was benefiting some countries, rather than all countries, and therefore the terms of the Outer Space Treaty were being applied in a way that did not reflect the spirit of its drafters. For the Bogota states, the argument was that a more just international order could not be achieved if the utilisation of space was reserved for only a select group of states, and that therefore the issues of the GEO and frequency spectrum should be resolved in accordance with the aims of the New International Economic Order. The framers of the declaration had no real hope of winning their argument, but sought rather to apply political pressure to the developed states that were monopolising the GEO.[30]

Some similar arguments emerged in relation to the 1979 Moon Treaty. Unlike the GEO and the frequency spectrum, the Moon has a tangible physical reality and in that sense is more understandable in relation to traditional concepts of territory. However, it is effectively inaccessible to virtually all states due to the economic and technological difficulties of travelling there. At the time of the 1979 treaty, therefore, it was not in any real sense a 'natural resource in restricted supply'. However, once the costs and difficulties of travel to the Moon were significantly reduced, there might be senses in which the Moon would represent a limited non-renewable resource, in respect of mining operations, for example.

Until the 1967 and 1979 treaties, the Moon was effectively 'res nullius', that is, owned by nobody, and not subject to exploitation by anybody. The 1967 treaty declared it not subject to national appropriation, but left it available for countries to exploit. For example, it was not clear as to what was the legal status of the lunar samples returned to Earth by the American Apollo astronauts. The United States sidestepped this difficulty by offering access to the samples for scientific research by any interested state.

In 1979, with the signing of the Moon Treaty, the Moon was declared to be not open to exploitation like the high seas, but rather part of the common heritage of mankind. Once the practical possibility of its exploitation became possible, an international regime would need to be established in order to ensure that such exploitation was on behalf of all mankind. This regime would amongst other things provide an 'equitable sharing by all state parties in the benefits derived from those resources, whereby the interests and needs of the developing countries, as well as the efforts of those countries which have contributed either directly or indirectly to the exploration of the Moon, shall be given equal consideration'.[31] Equitable in this case does not mean equal. Nevertheless, the United States and Soviet Union refused to sign the treaty. The United States in particular rejected it on the basis of the argument that unless states or companies had a right to the benefits derived from such exploitation, they would have no incentive to carry out the exploitation in the first place and therefore the lunar resources would always remain untouched and would not benefit humanity in any way. Thus, like the GEO and the frequency spectrum, the Moon has become part of the wider North–South debate on how the international community should best ensure 'equitable access' to the common heritage of mankind.

COPUOS has still to achieve a regime for the GEO that is acceptable to both the developed and developing nations. However, the UN's Unispace 82 Conference did produce conclusions declaring that the use of the GEO should be both equitable and flexible, and that it should take into account all relevant economic, technical and legal aspects.[32] In the same year the ITU, in response to the concerns of the developing countries, amended Article 33 (2) of its Convention to take into account the 'special needs of the developing countries and the geographical situation of the developing countries'. Access to the GEO and the radiofrequency spectrum would no longer depend entirely on the needs and technical facilities at the disposal of specific states. The phrase 'equitable access' continued to remain undefined, so that the 1982 amendment did not resolve all the issues. However, at the 1985 and 1988 ITU World Administrative Radio Conferences on the issue of the GEO, the developing states won their argument to some extent, with an ITU commitment to allow two GEO slots and associated frequencies to each state regardless of other considerations.[33]

Utilisation of the radiofrequency spectrum is a second area of dispute between the developed and developing nations in regard to space. Radio waves are a portion of the larger spectrum of electromagnetic waves. They are used for a wide variety of purposes, including telephones, am and fm radio, uhf and vhf television, air and maritime navigation, radio-astronomy, radar, meteorology and electronic mail. To communicate via radio waves the user must have exclusive access to the frequency over the geographical area determined by the distance the signal must travel to the receiver.

The bands below 30 MHz are particularly popular because they can be used to transmit signals over greater distances without having to use relay facilities. This is because the ionosphere bends the signals back towards the Earth's surface, allowing them to follow the curvature of the planet. However, these lower frequencies can accommodate fewer users without interfering with one another. Thus, with their limited capacity and greater popularity, the lower regions of the spectrum have become an especially scarce and contested resource, in need of international management. Developing countries pointed out that the most popular C and Ku frequency bands were cheaper to operate and less technically demanding than systems using higher frequencies. They were therefore particularly useful for the kind of remote area communications that offered the greatest developmental pay-off for the developing states.[34]

There are three possible principles of ownership that could be applied to the radio spectrum and the GEO. The most straightforward is when a resource is deemed to be owned by a state, such as oil reserves on a national territory, or fish stocks within a country's exclusive economic zone. A second possibility is that it is owned by no-one (res nullius). It is then not subject to regulation or management. However, some zones of this sort have been historically appropriated subsequently, as happened to parts of Antarctica for example. The third possibility is that the resource is deemed to be the common property of all (res communis). In this case it cannot be appropriated by individual states and is deemed to be part of the 'common heritage of mankind'. When a resource is deemed res communis, it is assumed that all states are entitled to participate in decisions regarding its use, and to share in the economic benefits of its exploitation.

Both the radio spectrum and GEO are deemed to be part of the common heritage of mankind. Radio waves move through space unimpeded by political boundaries, and it would be impractical for states to attempt to control airwaves over their territories. They can jam unwanted signals coming from outside their borders, but they cannot make constructive use of signals passing across their territories. But the issue of interference makes it essential for there to be an international regime regulating management of the airwaves.

The same is true of the GEO. It is an established principle of international law that Earth-orbiting satellites operate in outer space, beyond the atmosphere and the political rights conferred by the existence of sovereign airspace. They are therefore not subject to the effects of sovereign territoriality. The use of the GEO is subject to the provisions of the 1967 Outer Space Treaty, particularly Article I which decrees that use of space must be carried out 'for the benefit and in the interest of all countries', and Article II which declares that space is not subject to national appropriation 'by claim of national sovereignty, by means of use or occupation, or by any other means'.

International policymaking on telecommunications is made almost exclusively at conferences of the International Telecommunications Union. The ITU was founded in 1865 as the International Telegraph Union. It adopted its current name in 1932 and became a specialised agency of the United Nations in 1947. The ITU is the UN agency that allocates frequencies to particular uses and regions via the International Frequency Registration Board. It began regulating radio communications in 1906 and has been making frequency allocations since 1927. The radio regulations are effectively treaties and are subject to national ratification.

The ITU also deals with issues relating to the GEO, since telecommunications have been the predominant use for the GEO. The system of allocation is a complex one, but essentially operates on a 'first come, first served' basis.

When applications are made for new allocations, or the extension of old ones, the IFRB gives consideration to how the demand will affect the users already in existence, and whether their established use will be interfered with. The allocation will not be awarded if it would have a detrimental effect on already established actors. Once a claim has been fixed with the Master International Frequency Register, the rights awarded are inviolable as long as the owner operates according to the accepted rules. While this is a reasonable principle, which the IFRB is policing as a technical issue relating to efficient use, it also has significant political implications. The system, while on the surface a purely administrative exercise, is in effect distributive.[35] The 'already existing users' are invariably the developed states, who have been part of the IFRB system since it began making allocations in 1927. The aspirant users are usually developing states, who in practice are disadvantaged by a system of 'first come, first served' or 'squatters rights' that prevents the dominance of the developed states from being reduced.[36]

The difficulty for the developing nations in regard to the utilisation of outer space for development mirrors the general problems seen in the manner in which global economic arrangements similarly disadvantage the poorer countries. Likewise the fact that the developing nations are latecomers to the international community since the end of the colonial empires means that they have to fit into a system that was designed to suit the interests of the developed states. For Article I of the Outer Space Treaty to have genuine meaning, international cooperation in the space domain would have to be conducted in a way that gave the developing states preferential treatment. This would be the only way to allow all nations to benefit in a genuinely 'equitable' fashion.[37] The reality is that the existing system is one where developing countries were not equal at the start and the system has established a legal framework that essentially codifies this inequality, hindering the developing countries in their efforts to enjoy the benefits of space-related resources.

Article 33 of the IT Convention guaranteed 'equitable rights' only in accordance with 'the technical facilities at their disposal'. Although agreed to by developing states, this acceptance is now seen as a political mistake, since it substantially negates the spirit of the Outer Space Treaty and can be seen as allowing a 'res nullius' regime for space. The developed states countered by arguing that such concerns were unfounded, that the capacity of the GEO was more than adequate for the foreseeable future, and that subsequent constraints would be overcome by technological advances. The latter argument, while clearly holding some validity, was weakened by the fact that the developing states were the least able to invest in the technological advances that would be required.

The ITU as a body saw itself as an objective, decision-making organisation, and believed that its decisions were taken on the basis of purely scientific and technical considerations. During the 1970s, however, it was steadily drawn into the larger confrontation between the developed 'North' and the less developed 'South', which swirled around the debates over the desirability of a 'new international economic order'. 'Third World' states began to question the justice of ITU policy. At the time of the 1979 World Administrative Radio Conference, for example, it was noted that 90 per cent of the radio spectrum was controlled by countries with only 10 per cent of the world's population. The new radicalism reflected the changing make-up of the international community. The 1960s had seen a large increase in the number of UN members, almost all of whom were post-colonial developing countries.

Many of these new states' criticisms were directed at the 'first-come, first served' policy used by the ITU for allocating GEO slots and radio frequencies. This, they argued, gave the developed world a permanent hold on the best parts of these limited natural resources. Hulsroj has compared the situation to that of the frontier experience in the 'Wild West', arguing that 'space in the sixties was a new frontier, but cultivation has long since been afoot in the part of space closest to Earth, and particularly in the geostationary orbit. Yet, we are still stuck with the equivalents of the law of the fastest draw'.[38] The developed states countered by arguing that the ITU policies had the effect of promoting the most 'efficient' use of the spectrum and orbit, thereby expanding capacity to transfer information, and that the only way to achieve this was to give the capacity to the most technologically advanced states.

For example, in the arguments over the popular and congested high frequency (HF) region of the radio spectrum, the developed states' solution was to shift domestic transmissions to microwave and cable, thereby freeing up more HF capacity, but at great cost in terms of the required capital investment. Some developed states particularly desired to free up the short-wave broadcasting for their long-distance propaganda stations such as Voice Of America, Radio Moscow and the BBC World Service. The developing countries argued that this was too expensive a solution given their economic weaknesses and proposed instead a solution involving the transfer of existing frequencies in the popular 'C' band, to specific countries in the developing world, with the more technologically developed states focussing their use on the 'Ku' and 'Ka' bands. The expenses involved in developing and operating systems installed at the higher frequencies would then fall on those countries best equipped, technologically and financially, to cope with them. The developing countries in addition had no desire to be exposed to even more foreign propaganda. However, the proposal was rejected by the developed states, who were anxious to protect their investments in the existing technology.[39] This was despite the fact that prognoses for the later decades of the century indicated that fibre-optic cable networks were likely to take a significant portion of communication traffic in the developed states and that the Ka band might be more cost-effective in the longer term.[40]

Developing countries argued that the logic of the Outer Space Treaty insistence that space should benefit all states 'irrespective of their degree of economic or social development' was that richer, more technologically advanced states should help them acquire the advanced technology needed to exploit the information technology revolution. Developed states such as the USA, in contrast, argued that the terms of the Outer Space Treaty mean only that states have a limited responsibility not to misuse space in a way that would reduce its value for the space activities of other countries.

The resolution of these issues fell upon the ITU's World Administrative Radio Conferences (WARCs). The 1979 WARC resolved to guarantee equitable access to the GEO and frequency spectrum.[41] By the time of the 1985 WARC two possible methods were identified. The first was the Allotment Plan. Under this scheme it was hoped to furnish all states with at least one allotment in the GEO for systems providing national services. The plan takes account of the current and future needs of the developing countries. Not every member of the ITU has the current need or ability to use the GEO, but the possibility that a position may be required in the future is recognised and given priority. The geographical position of each state is taken into account so that the allotment they are given is one that would be of practical value to the state concerned.

The second method was Improved Procedures Planning and was intended to address the other issues in the FSS service. It dealt with the congested radio frequency bands. Due to heavy demand, the efficient, economical and flexible use of these frequencies was considered along with the existing need of each country to use them. The special needs of the developing world were given much less priority in this planning method, which instead emphasised the question of which nations could currently use them most efficiently.

The use of two separate methodologies was designed to meet the ITU's conflicting goals of providing 'equitable access' and maximising the efficient use of the resources. Eleven principles were adopted that were intended to apply to both methods, and that attempted to balance these obligations in a way that met the legitimate concerns of the developing states. The ITU, as a UN body, is aware of the paradox of space technology, that its potential benefits are greatest for the countries that are least able to take advantage of it. The UNISPACE 82 Conference stressed the political implications of this by calling upon the developed nations to 'promote the wider exploitation of space technology by developing countries. Space technology can be a powerful tool to accelerate national development: it provides a way of leap-frogging over obsolete technologies and getting away from percolation and trickle-down methods of development for which developing countries do not have the time.'[42]

UNISPACE 82 provided a forum for the developing countries to express their unhappiness at the existing space regime.[43] Delegates from the poorer states complained that data exchanges with developed states were in practice infrequent because of 'national security' concerns on the part of the developed nations. The developed states treated cooperation as little more than public relations exercises, the industrialised world did not recognise developing world scientists as worthy contributors, and any cooperative programmes suffered from a lack of realistic funding.[44]

The problems created by 'national security' concerns are not specific to the developed world, however. At the time of the 1999 UNISPACE III, a number of different regions in the developing world emphasised the need for greater international cooperation, coordination and data access. But there was evidence that many of the restrictions on data sharing came from within the developing countries themselves, because of fears that regional satellite imagery, acquired for reasons such as sustainable agricultural development, could also be used for military purposes.[45]

Developing states in the 1970s and 1980s called for a New International Information Order and were particularly sensitive with regard to remote sensing data acquired without the specific permission of national governments. This was seen as 'a threat to their national sovereignty and their "sovereign right" to control information about themselves and their resources'.[46] The focus on issues of sovereignty, and the impact of space technology into questioning its continuing defensibility in this debate is further evidence of the way in which the space age saw key modernist concepts of international politics significantly undermined or seriously questioned by the new realities created by space technology and its application.

The North–South division in terms of space utilisation was also apparent within INTELSAT, the global communications satellite cooperative. INTELSAT was created in 1964 and includes governments and business enterprises in its membership. It is dominated by the United States through the hegemonic position of COMSAT within the organisation. Because it was responsible for creating the organisation and pays the largest financial contribution, the United States once controlled 53 per cent of the voting power in the organisation. Despite membership having increased to 110 countries, the USA still has 43 per cent of the vote and dominates the organisation.[47] Not surprisingly, developing countries felt that the organisation pursued policies that benefited the United States, but were detrimental to their own interests. For example, INTELSAT chose to employ satellite space segments designed to serve high volume, public switched traffic best, particularly in the North Atlantic market. As a result, Earth station investment and operating costs averaged twice as much for developing and newly industrialising countries as it did for the developed industrialised states. Opting for a different technology might have raised the costs of the space segment, but would have sharply reduced earth station costs, especially for small users, thereby producing substantially lower total system costs for developing states. From the developing states' perspective, therefore, INTELSAT consistently opted for solutions that benefited the developed world, but pushed a heavier burden on the developing countries than was necessary. From the developed states' perspective INTELSAT was seen as operating satisfactorily 'from both the technical and legal points of view'.[48]

During the 1990s developing nations began similar moves within the legal subcommittee of the UN Committee on the peaceful uses of Outer Space (COPUOS). The rapid growth of the internet had made it increasingly obvious that it was critical that developing countries participate as fully as possible in the telecommunications revolution if they were to reap the benefits of the information age.[49] These were based on the wording of Article I of the Outer Space Treaty, which declared that space activities should be for the benefit of all countries, 'irrespective of their degree of economic or scientific development'.[50] The UN responded to these pressures with General Assembly Resolution 51/122.[51] The resolution reaffirmed the terms of Article 1 of the Outer Space Treaty, and stated, among other things, that space programmes should encourage international cooperation and that 'particular attention should be given to the benefit and the interest of developing countries and countries with incipient space programmes'.[52]

It is important to recognise that the 'developing states' is a rather vague and unwieldy category,[53] and in fact there are wide variations between states in this category, as well as a degree of movement of states out of the developing and into the developed category. In the past two decades some developing countries such as China, India and Brazil have emerged as major space powers in their own right, so that their interests and outlooks do not necessarily coincide with other 'developing' nations.[54]

Conclusion

The debate about the place of development within international relations raises important questions that are relevant to an understanding of the international politics of space. Moreover, the study of the latter as a field of inquiry requires a focus on primarily national space programmes, while at the same time making clear the limitations for human development of a global political system structured around nation-states. The utilisation and exploration of space is a naturally 'federative' activity, encouraging and requiring international cooperation, and, as Dower has likewise noted, 'there is no reason at all why the unit of development thinking must be the nation state'.[54] One effective way of changing thinking in regard to these areas would be to apply the broader concept of 'security' to allow the human security aspect of development to place it higher on an international agenda that has traditionally prioritised security issues above all others.

Chapter 9

India

Security through space

India's space programme is in some ways the most cost-effective and successful space programme in the world. The technological feats achieved by the Indian Space Research Organisation (ISRO) are dramatic achievements for a developing country that at the end of the Cold War was still one of the poorest in the world. Even more impressive than that, however, is the consistent way in which India has sought to use space as a crucial mechanism for lifting India's people out of poverty through education and social and economic programmes.

Independent India's first Prime Minister, Jawaharlal Nehru, declared that 'science alone can solve the problems of hunger, insanitation and illiteracy'.[1] This sentiment was echoed two decades later by Prime Minister Rajiv Gandhi who declared in 1989 that 'we must also remember that technological backwardness also leads to subjugation'.[2] Similarly, inaugurating the satellite-linked Village Resource Centres in 2004, the current Prime Minister Manmohan Singh recalled Nehru's emphasis on science and technology and declared that of all the institutions created since independence, none had brought India greater credit or worked so consistently to be socially relevant, as the Indian Space Programme.[3]

During the latter part of the Cold War, analysis of the Indian space programme by international relations scholars tended to raise alarmist concerns about the programme, by highlighting the alleged dichotomy between the development goals that Indian governments historically used to justify the programme, and the obvious military rationale that lies behind the country's development of both launcher and satellite technology.[4] As with parallel criticisms of India's nuclear programme, there was a distinct element of hypocrisy in critiques that were sanguine about the American record in this area, but deemed a comparable effort by the world's largest democracy to be a potentially destabilising factor in international relations.

Subsequent development histories have made it clear that the Indian space and missile development programmes are distinct enterprises, notwithstanding their common roots, and the technologies employed are far from interoperable. In any case, a better way of understanding the Indian programme is not by seeing the development and military rationales as distinct objectives in a managed tension with each other, but rather to see them as parallel components of a coherent long-term *security* policy.

A feature of international relations theorising since the early 1980s has been the broadening of the concept of 'security' to move beyond a limited focus on the military and state dimensions of security to embrace other dimensions, notably economic, environmental, societal and political dimensions.[5] Seen in this light, the Indian space programme is distinctive and coherent in the way that it simultaneously addresses the requirements of the Indian people and state in all these dimensions.

India's Department of Space makes clear this continuum in its programme rationale. In its 2006 'Citizens Charter', the DOS declared that its objective is to 'assist in all round development of the nation' and that this embraces, *inter alia*, providing 'satellite imagery required for the developmental and security needs of the nation'.[6]

The first Director of ISRO, Dr Vikram Sarabhai, was a committed proponent of the view that the essential rationale for an Indian programme was to shorten the time that it would take to eliminate the fundamental poverty in which the vast majority of India's people lived. But given the expense and technological obstacles associated with a space programme, he was deeply aware of the need to demonstrate its developmental function at all times to what was likely to be otherwise a sceptical Indian political élite and population. For Sarabhai and his successors, the programme's purpose was to be 'second to none in the application of advanced technologies to the real problems of man and society, which we find in our country'.[7]

The subsequent 40 years of the programme have vindicated this vision, without reducing the need to engage continuously with public and political opinion. As a developing country, India's space programme needs to be domestically justified in terms of a clear development rationale in a way which is not the case with the programmes of more developed countries such as France and Japan, which have focussed on goals of scientific discovery and financial reward. In 2005, for example, 76 Parliamentary questions about the space programme were tabled in the two Houses of India's Parliament.[8] Successive governments have argued that the solutions for many of India's development problems may well be found in space technology,[9] but all have remained sensitive to the accusation that India should not be devoting precious resources to space technology, when there are so many problems on the ground demanding government attention. Governments therefore make continuous efforts to identify the Indian population with the content and achievements of the space programme. The Department of Space runs roadshows and outreach programmes throughout India designed to maintain public interest in, and support for, the space programme. In addition, the space programme features in school curricula at all levels.[10] These efforts appear to have been successful in maintaining high levels of popular approval of the space programme.

The developmental ideology has become firmly embedded in the self-perception of the space programme, allowing Indian observers to declare confidently that 'while nations with advanced space technologies were thinking in terms of exploiting space for civil and military purposes, India's goal was of self-reliant use of space technology for national development'.[11] This was crucial for a country that even at the end of the Cold War had 75 per cent of its population living in the countryside, with very high levels of illiteracy and an average per capita income of $170, making it one of the 20 poorest countries in the world.[12] India sees its commitment as clearly justified by the fact that for India, 'when compared to conventional methods, satellite remote sensing methods are cheaper and faster at least by a factor of 2–3, and more in some cases'.[13]

Though the contemporary space programme is in dramatic contrast to Indian underdevelopment and a striking achievement for a newly independent country, India in fact has a long historical legacy in terms of interest in astronomy and rocketry. The earliest written account of anything resembling a space launch can be found in ancient Sanskrit texts such as the *Rig Veda* and the *Mahbharatha*, which speak of a vessel called the *Vimana* ascending to heaven. The *Vimana* is described as giving 'forth a fierce glow, the whole sky was ablaze, it made a roaring like thunderclouds' and then took off.[14] When India's first satellite was launched in 1975, it was named *Aryabhata* after one of India's (and the world's) greatest astronomers and mathematicians. Aryabhata (476–550 AD) in the year 500 produced a theory arguing that the Earth is a sphere, that it spins on its axis, and that the planet's motions are relative to the sun. These ideas predated those of Copernicus by a thousand years. Sanskrit texts dealing with astronomy date back to 1350 BC, and some of these texts correctly identified the stars as distant suns.

Indian armies were using rockets for military purposes as early as the eighteenth century. The Sultan of Mysore had rocket companies attached to each brigade of his army, 27 in all. At the Battle of Pollilur in 1780 the Mysore army defeated a British army, after rockets destroyed the British ammunition wagons. Rockets were also used against British armies at the battles of Seringapatam (1792) and Srirangapattana (1799). The British were so impressed by these rockets that large numbers were sent back to Britain and reverse-engineered by Sir William Congreve, leading to the creation of Rocket Troops in the Royal Artillery.

This historical legacy is important. Unlike some of the post-colonial states that became independent after 1945, India is not an invented nationality with borders created partly by colonialism. India's people take pride in being the inheritors of one of the world's great civilisations, with a history of cultural and scientific achievement that is millennia old. For a large, proud nation like India the space programme, like the armed forces and the national airline is a 'flag-carrier', an emblem of successful nationhood. In developing space launch vehicles and ballistic missiles, India has succeeded in a 'demanding test of national strength that few third world countries can hope to pass'.[15] Indeed India's achievements in her space programme puts her in the second tier of space-faring states alongside countries like France, China and Japan, with only Russia and the United States possessing more advanced capabilities.[16]

India's programme began within a few years of the launch of Sputnik. That it began at such an early stage, so soon after the initiation of space programmes by the superpowers, was a remarkable decision for a developing country, contrasting, for example, with Britain's reluctance to pursue a national programme at this point, because of the risks and expense involved. The Nehru government's decision to create an Indian space programme reflected Nehru's belief in the power of science to bring development, but even so represented 'an act of extraordinary foresight and courage' given the novelty and complexity of space technology at this time and the grim realities of India's economic situation.[17] The Indian National Committee for Space Research (INCOSPAR), was established in 1962, with a remit to advise the government on space policy and to foster international cooperation in this field. After several years of developing sounding rocket technology with help from Britain and France, India created ISRO, the Indian Space Research Organisation, in 1969.

The Indian space programme came into formal existence in 1972 when the government established the Space Commission, whose mandate was to review the development and application of space technology and space sciences in terms of promoting India's national development. A new Department of Space (DOS) was given the responsibility for implementing policy in space applications, space technology and space science through ISRO. The creation of the Space Commission and the Department of Space symbolised the transfer of responsibility for space research directly into the hands of the government. The DOS was now responsible for policy implementation in all areas of space activity. Crucially, the Department of Space is directly answerable to the Prime Minister.

The government resolution which established the Space Commission declared that India attached the highest priority to the development of space science and its applications, and justified the new organisational structure on the grounds of 'the sophistication of this technology, the newness of the field, the strategic nature of its development and the many areas in which it has applications'.[18] This view significantly predated the United Nations' recognition of the potential space technology had for developing countries, when it noted that 'space technology can be a powerful tool to accelerate national development; it provides a way of leapfrogging over obsolete technologies and getting away from percolation and trickle down models of development for which developing countries do not have the time'.[19]

A number of driving factors lay behind the momentum to establish an Indian space programme. As a recently independent state, with a population still smarting from the memory of the colonial experience, India sought to achieve genuine technological independence from the developed world, and to be free from any constraints that would accompany a dependent interaction with the technologically advanced nations. The initial rationale of the space programme was to develop mass communication, particularly television, throughout India, and to use Earth observation satellites to monitor and manage the country's natural resources, and ensure they were all put to productive use.[20] Second, India sought the prestige of belonging to the élite grouping of the world's 'space powers'; lastly, she sought the economic benefits that might come from the exploitation of space. For successive governments, the space programme offered the possibility of allowing a developing country to bypass the intermediate technology stage and move directly into the high-technology era.[21]

Proponents of the space programme argue that the emphasis on indigenous development is critical. The technology boost provided by the programme has an impact well beyond space activity itself and, as was the case in the United States, has the potential to accelerate progress in a wide range of technologies. In the early stages of the programme, however, India was necessarily dependent on cooperation with other countries to acquire initial technological and engineering skills in an unfamiliar realm. The risk of dependence was reduced by a conscious effort to cooperate with a number of different countries simultaneously.

The first launch from India's Thumba equatorial launch site took place in November 1963. From the outset, the Indian effort was marked by an openness to international cooperation. The 1963 launch was part of a UN sponsored international effort. Between 1963 and 1975 some 350 rockets of American, British, French and Soviet design were launched from Thumba. India's first significant indigenous rocket, Rohini-75 was launched on 20 November 1967. It was a small sounding rocket with a diameter of only 75 cm, but in launching it India proved that it had mastered the basics of modern space rocketry. Alongside the Rohini programme, India was cooperating with France to manufacture the two-stage Centaure rocket. The experience of codeveloping Centaure was crucial in terms of India's national programme. During this period also, Hideo Itokawa, the 'father' of the Japanese space programme was employed as an adviser to the Indian programme, on the invitation of Prime Minister Indira Gandhi.

India's first satellite, Aryabhata, was launched on 19 April 1975 on a Soviet rocket from a Soviet launch site. While this clearly demonstrated a reliance on a major external power, the cooperation with the Soviet Union was also a way of balancing the earlier cooperation with the United States, an important consideration given that India was one of the leading states in the non-Aligned Movement. The satellite itself, however, was entirely Indian, and was a test bed for Indian satellite technology. The cooperation with the Soviet Union, particularly in the areas of system manufacture and satellite monitoring, was to be crucial in the development of an indigenous Indian solid-fuel rocket later in the decade.[22] In 1980 India became only the sixth state to launch successfully a satellite using its own launch vehicle.[23] The following year India gave a further demonstration of her rapidly developing capabilities with the launching of an Indian satellite into geostationary earth orbit (GEO) by the European Space Agency in June 1981. India became only the fifth country to master the demanding technology required for this, after the superpowers, France and Canada.[24]

For India, entry into the satellite era held the promise of exploiting space for the purpose of developing India's resources on Earth. Remote sensing and communications benefits were identified at an early stage. These included the acquisition of data on soil and water resources, crop surveys for both production levels and disease detection, weather forecasting – critical to a country with a tropical climate marked by extremes of weather – telecommunications and, most notably, the use of satellites for rural development via education. Sixty per cent of India's population still live in small rural communities, and the use of satellite technology has been crucial in the establishment of a national telecommunications network linking these communities to the outreach of Indian government development policy. These capabilities were seen as being not a luxury, but a necessity for a developing country.[25] They produced visible benefits fairly quickly. In 1988 the IRS-1 satellite enabled the Geographical Survey of India to locate a 60 kilometre (37 mile) extension of a known fault line containing deposits of lead and zinc in Andra Pradesh. Satellites have made an effective contribution to pest control policy, allowing up to 25 days warning of the movement of locust swarms by tracking the winds that carry them. Satellites have also been used against other pests such as cotton flies and the brown plant-hopper, whose infestations can be detected from orbit.[26] Examples of the use of space technology in relation to natural disasters include the Cyclone Warning Dissemination System, which provides meteorological data and is linked to over 250 disaster-warning receivers at stations on the cyclone-prone eastern coast of India. The satellite system directly triggers klaxon warnings and recorded messages in local languages when cyclones are imminent. Whereas during the 1970s coastal cyclones caused thousands of deaths in India, few deaths were reported once the satellite-based meteorological and early-warning systems became operational.[27] The Department of Space has also used satellite monitoring of Himalayan snow cover levels since 1994 in order to predict the volume of snow that will melt, and the implications of this for flooding at lower altitudes.[28]

The use of satellites for educational purposes was seen as being particularly important. India's SITE (Satellite Instructional Television Experiment) programme was inaugurated on 1 August 1975 using the American-launched ATS-6 satellite.[29] ATS-6 operated on 20 frequencies with a direct broadcast capability to 2,400 community receiving sets. Four hundred receivers were located in each of six Indian sites – Andra Pradesh, Karnataka, Orissa, Madhya Pradesh, Rajasthan and Bihar. In addition to villages directly receiving programmes by satellite, others participated through rebroadcast by television stations. The programmes were beamed to an estimated 2,400 villages and watched by five million people.[30] The objective of the SITE programme was to test the effectiveness of television as a medium of communication for national development in backward rural areas. The programme concentrated on general education, agriculture, health and family planning. Emphasis was placed upon the prevention and curative aspects of general health, maternal and child health care, nutrition and birth control.[31] The agricultural education emphasised sources of supply of agricultural inputs such as seeds, fertilisers, implements, pesticides, credit, pest and disease control and weather forecasts. The steady development of this programme resulted in the 'evolution of a unique satellite concept, Gramsat (gram meaning village), tailored to disseminating culture-specific knowledge on health, hygiene, the environment, family planning, better agricultural practices etc., to the vast and diverse rural India'.[32]

The benefits which were gained through access to satellite technology prompted India to undertake an expansion of her capabilities and lead to the launch of her first domestically produced communications satellite in 1992. The Insat 2A dual payload satellite was the result of over a decade's research at a cost of $250 million.[33] It fulfilled both telecommunication and meteorological functions thereby ending India's reliance on technology purchased overseas in this field. At the end of the 1970s, only 30 per cent of India's population had access to television, and the Insat series was developed in an attempt to raise this figure to 75 per cent. In the event, by the turn of the century, 90 per cent of the population were receiving television broadcasts and there were 700 television stations in operation.[34] India's seven communications satellites constitute the largest civilian system in the Asia-Pacific region.[35]

From the outset a major objective of ISRO was to develop the application of remote-sensing techniques to survey India's natural resources, particularly agriculture.[36] An early experimental project, the Agricultural Resource Inventory and Survey Experiment (ARISE) was used to collect data on agricultural resources by remote sensors and correlate them to pre-established ground-truth data. It was determined that space photography could be used to determine soil types and soil moisture levels, and, using false colour contrast, to identify different crops. Infrared photography clearly detected water sources such as streams, canals, wells and water-logged areas.

Indirectly, it was possible to interpret underground water patterns. Infrared satellite photography was also used to monitor the spread of disease in crops, alerting farmers even before they themselves were aware of the problem.[37] These techniques were developed through the 1970s and 1980s to produce an integrated natural resources survey programme coordinated by India's National Remote Sensing Agency. The CAPE Programme (Crop Acreage and Production Estimation) uses satellite data to allow the Department of Agriculture to make pre-harvest estimates for major food crops as well as cotton.

Each year India loses billions of rupees to damage done to agriculture by pests, floods and droughts, in addition to the famine risks produced by these dangers. Remote sensing satellite technology was a direct response to these problems, allowing prediction and monitoring of major agricultural problems. The National Natural Resources Management System was established to provide data relevant to drought monitoring, flood prevention, mineral development, land mapping and water resources management.[38] Eleven Indian states currently receive fortnightly bulletins on crop conditions and drought dangers. In 1987 the Indian government committed itself to ensuring that every village had access to 40 litres per person of clean drinking water per day. To help achieve these goals the Department of Space was called upon to use its satellites to produce maps that could help locate sources of underground water. One of the effects of the orbiting of India's IRS-1A satellite launched by the Soviet Union in 1988 was 'an increase in the success rate of bore-drilling from 45% to 90%'.[39] Landslide Hazard Zonation mapping from orbit is used by India to determine landslide risks for all major pilgrim and tourist routes in the Himalayas, Uttaranchal and Himanchal.

India's advances in the field of remote-sensing techniques have been such that with the launch of the IRS-1B satellite on 29 August 1991 she entered into the business of marketing the results of this technology. IRS-1B, like its predecessor IRS-1A, was sent into a 900 km polar orbit using a Soviet launcher, having been designed and assembled within India. The imagery resulting from the IRS-1B may be received by any ground station currently capable of receiving US LANDSAT or French SPOT satellite data with only a few minor technical alterations.[40] The data available will be of use in those areas which have proved to be of particular interest to India itself, including meteorology, agriculture, mineral exploration and urban and rural development. India's National Remote Sensing Agency receives data not only from India's own Insat series satellites, but also from the American LANDSAT and European ERS satellites.[41] By the beginning of the twenty-first century India was operating six remote sensing satellites, the largest concentration in the world.[42]

Remote observation has benefited India more than it would have a developed country. By 1970 barely 20 per cent of India's agricultural land had been surveyed in detail using traditional methods, whereas in developed states such as Britain, full surveys had long been in existence. An early achievement of the space programme was the creation of a national series of 1:250,000 scale maps for agricultural development purposes. These maps helped identify potentially productive arable land, and also allowed the siting of new roads, so that as far as was possible they did not use up actual or potential agriculturally productive land, as well as being in better conformity with the terrain.[43] The IRS-1A satellite provided data indicating that India's forest cover was disappearing rapidly. IRS data has also been used to predict the location of fish concentrations off the coasts of Gujarat, Maharashtra and Andhra Pradesh, producing dramatic increases in catches. OCEANSAT, launched in 1999, monitors the surface temperature and chlorophyll content of the sea and detects upwells of cold, nutrient-rich water. Such upwells attract large numbers of fish and the data is passed on to 200 coastal centres allowing fishing fleets to converge on the areas. Fish catches have doubled in the past ten years.[44]

While it is difficult to place a precise monetary figure on the contribution of the space programme to India's GDP or sustainable development goals, it seems reasonable to apply the assumptions used by the United States government that investment in information sources such as remote sensing generates economic benefits equivalent to between five and ten times the original investment.[45] By this standard the space programme has made a huge contribution to India's economic and human security development in the past 25 years.

Much of this work was carried out by India's National Remote Sensing Agency (NRSA), an autonomous organisation within the Department of Space. Its role is to enable India to exploit the benefits of remote sensing in identifying and surveying natural resources. Overall strategy for matching space technology with development goals is provided by the Development and Educational Communication Unit (DECU), at Ahmedabad, which focuses on 'the conception, definition, planning, implementation and socio-economic evaluation and developmental space applications'.[46] At all levels therefore the Indian space programme is structured so that it can effectively implement national development goals, and its organisation clearly reflects the nature of those goals.

As a genuinely non-aligned state during the 1970s and 1980s India had friendly and cooperative relationships with both the United States and the Soviet Union. This was a great advantage to its developing space programme, which benefited from launch facilities and technical cooperation with both superpowers. Though the Indian space programme clearly benefited from this collaboration, it remains a sensitive issue within India which prides itself on the degree of self-reliance and technological initiative demonstrated by the rocket and space development programmes. Some observers of the Indian programme have been sceptical about the degree of genuine self-development the programme indicates given the continuing reliance on external partners for key areas of technological collaboration.[47]

International cooperation remains an important dimension of India's programme. For example, India's National Remote Sensing Agency receives data from the US LANDSAT 5 and European ERS-2 as well as the Indian constellation of satellites.[48] This has the practical advantage of gaining access to additional data, but it also has the effect of linking India with the activities of other space organisations. Moreover, there are dimensions to India's security problems such as climate control, large-scale environmental degradation and natural resource depletion that Indian governments have always felt were most appropriately addressed through collective action by the international community, rather than through specifically national policies. Such issues need to be addressed 'from a global, rather than a national point of view'.[49]

It has been argued that India's space programme has two objectives. Firstly to assist in India's social and economic development and secondly to provide India with diplomatic leverage in terms of contributing to the development of international space law and the maintenance of space as an area of peaceful exploration and exploitation.[50] Nevertheless, there is also a third, less publicised objective, the military one.

The military dimension

The military implications of India's space programme have been noted by a number of writers.[51] Awareness of external concern on this issue led Indian ministers to state publicly that no plans existed to use the launcher programme to acquire an IRBM capability.[52] Government ministers, such as Dr Satish Dhawan, Chairman of ISRO and Secretary of the Department of Space, argued in 1980 that the Indian space programme would have been unnecessarily slowed if it had been a combined military and civil space programme, and that the cooperation with other states which had benefited ISRO would have been impossible if India had possessed an obvious long-range ballistic missile programme.[53] This was a somewhat disingenuous claim, since there *was* in fact such a ballistic missile development programme, but officials were correct in stating that it was not being carried out using ISRO technology. The missile programme was a defence ministry responsibility, utilising the Agni missile technology.

India's geopolitical situation, bordered by China and Pakistan, both of whom she has fought wars with since independence, naturally encouraged the desire to maximise the military and political effectiveness of her armed forces, and therefore to use its potential to enhance its military capability through the 'passive' use of satellites as a force-multiplier. The Earth-observation satellites India has launched can be used for military reconnaissance as well as development purposes, monitoring troop movements and buildups, major military facilities and weapons development sites. Those with infrared systems technology also allow such observation to continue in the hours of darkness.[54] The C and D series IRS satellites can produce 5.8 metre resolution imagery that has clear military utility.[55] While it can be used to coordinate Indian military planning and to help target Indian weaponry in wartime, it is also true that reliable reconnaissance capabilities can act as a confidence-building measure between states suspicious of each other's intentions and activities, such as India and Pakistan, or India and China. Indian security specialists see satellite reconnaissance capabilities as being key both to monitoring military security *vis-à-vis* both China and Pakistan, and as being central to the achievement of future arms control and disarmament agreements.[56] If India should seek to scale down or eliminate the military confrontation with those states through arms control treaties, the remote-sensing satellites would be crucial to the successful verification of such agreements. Satellite remote sensing, like most satellite technology, is effectively dual-capable. When the Technology Experimental Satellite was launched in 2001 with a one-metre resolution capability, its military potential was clear. ISRO Chairman Dr Kasturianga noted that 'all Earth observation satellites look at the Earth. Whether you call it Earth observation or spying, it is a matter of interpretation. All I can say is that we have built this particular payload, which is used for imaging, as a forerunner to an advanced imaging system of a high resolution type'.[57]

Military communications satellites have a major potential for India's armed forces, particularly for the Indian navy. Moreover, with deployments scattered over a subcontinent, the Indian army (the world's fourth largest) and air force (the fifth largest in the world) would derive tremendous benefit from a satellite-based communications network. The navy would also benefit from maritime observation capability. India launched its first ocean observation satellite (IRS-P4 or *Oceansat*), on 25 May 1999, in the presence of the Prime Minister. It monitors the Indian Ocean and Bay of Bengal in particular. The launch had been 'delayed by American sanctions flowing from India's nuclear tests of 1998', which meant that the American-built ocean colour monitor had to be replaced by a German equivalent.[58]

Indian strategic analysts have argued that the benefits of acquiring high-technology weapons are the same for developing states as they are for those in the developed world. They confer a military advantage over an adversary.[59] It is not just this which motivates developing states like India, however, there is also the belief, reinforced by the coalition victory over Iraq in 1991, that developing states are highly vulnerable to intervention by developed states operating with all the benefits of the most advanced military technology.[60]

India's expertise in the field of guided missile technology grew steadily during the 1980s, as the result of a research programme initiated under Prime Minister Indira Gandhi in 1983, the Integrated Guided Missile Development Programme (IGMDP). This programme produced the short-range missile TRISHUL, a surface-to-air missile and the PRITHYI, a surface-to-surface missile with a range of 160 miles.[61]

It was the IRBM potential of India's launcher technology that generated the most controversy, however. In 1979 Professor Satish Dhawan, Chairman of the Space Commission and Secretary to the government in the Department of Space, declared that the SLV-3 rocket could be converted into an IRBM with a 1500 km range.[62] Support for the IRBM option clearly existed within the Indian military with one military publication calling in November 1981 for India to acquire an 'adequate capability for strategic long-range strike in the form of MRBM/IRBMs equipped with nuclear warheads'.[63] In 1990 the Director of the Indian Institute for Defence Studies and Analysis called for the creation of an Indian 'Strategic Air Command'.[64] The missile development programme was clearly related to the acquisition of nuclear weapons technology.[65]

Such a capability would require both technological capability and political will. India crossed the nuclear threshold in 1998 with a series of weapons tests. With regard to launcher and re-entry vehicle technology, an ablative heat shield for an RV has been worked on, while the long-range Agni variant is being developed as the delivery vehicle. The Agni was first tested in 1989 and has now been developed to IRBM capability with a range of over 2000 km and is capable of delivering a nuclear warhead. The Agni uses a two-stage missile system, the first stage of which uses the same solid-fuel booster as the civilian SLV-3.[66] For the missile to deliver warheads (of any kind) to target, sophisticated guidance mechanisms are required. By manoeuvring satellites to geosynchronous orbit India has shown that she is capable of developing such guidance technology. In 2001 India launched three satellites into orbit on the sixth PSLV flight. It had carried out the same feat in 1999. Such technology is capable of being adapted for MRV capability if desired.

Should this happen, the IRBM force would be oriented towards deterring Pakistan or China, or both. India herself has never officially designated either of these countries as the object of a policy of deterrence, but has rather stressed the general deterrent effect of her missile forces.[67] India's military planning has been aptly characterised as being based on the principle of 'keeping one step ahead of Pakistan and at par with China'.[68] Pakistan is India's 'traditional' enemy given that relations have been cool since independence in 1947 and have been punctuated by a series of wars, as well as continuing tension and sporadic fighting over the disputed territory of Kashmir. Although India has a marked superiority in conventional forces, the acquisition of nuclear capability by Pakistan in 1998 changed the strategic calculus, making an IRBM capability clearly desirable, as well as dramatically raising the stakes involved in any conflict.

The Chinese threat is more potent. China emerged as a major security threat when it invaded India in 1962 in a dispute over territory where China has remained in occupation since. Until then India had seen China as a political rival, but not as a military threat. The emergence of the Chinese threat significantly increased India's feelings of insecurity.[69] To a significant extent also the Pakistani and Chinese threats coalesced in 1965, when China issued a military ultimatum to India during her second war with Pakistan.[70] China possesses a substantial force of medium and long-range ballistic nuclear weapons and while these are orientated towards the deterrence of Russia, they are available for use, if required, against India. It has been suggested by some military analysts that the improvement in relations between Russia and China since 1985 may have led to an increase in the Chinese missile threat to India.[71] India has long believed that it is China, not Pakistan, against which India's capabilities and influence should be measured. India's missile build-up during the 1980s reflected the judgement of Indian strategists that 'the appropriate and logical point of reference to define India's strategies would be in relation to the People's Republic of China'.[72] India and China are political rivals, with a pattern of opposing regional alliances (for example, China with Burma, India with Vietnam) as well as a number of serious bilateral conflicts of interest. Though both emphasise development goals their relationship is inevitably one marked by suspicion and conflict. In the event of a sustained and dramatic deterioration in Sino–Indian relations, India might feel her security enhanced by being in possession of an IRBM force, carrying a nuclear deterrent. Unlike the case with Pakistan, if India were to attempt to threaten China with a nuclear deterrent, she could not use manned aircraft to deliver the weapons since China's cities and major industries lie beyond the operational range of Indian aircraft. Given that India's major launch site is 3,000 miles from the Chinese border, a missile of some 6,000 mile range would be needed – an ICBM rather than an IRBM. Apart from the Agni development, the advent of the GSLV has led some American observers to declare that India now has the capability to deploy an ICBM.[73] Any such effort would be complicated by the fact that the third stage of GSLV uses a cryogenic stage which is procured form Glavcosmos in Russia.[74]

The issue of self-reliance is a crucial one for India. India took particular pride in the development of Agni because it was argued that it was achieved entirely through India's own efforts, and because of this the missile was not covered by the inspection process of the 1987 Missile Technology Control Regime. However, a report published in the United States demonstrated that key areas of both the SLV-3 and Agni rocket systems were derived from technologies acquired during the 1960s and 1970s from abroad, in particular from the United States and the Federal Republic of Germany.[75] In particular it has been alleged that the Federal Republic of Germany may have played a critical role in the development of the Agni missile.[76] France has also been of considerable benefit to India's programme to date, cooperating in coproducing the Centaure rocket in the 1960s, helping to develop an Indian version of the French Viking engine in the 1970s and offering to sell cryogenic engine technology in the late 1980s.

While external assistance helped the early development of the programme, India has now achieved a level of self-sufficiency comparable with most other national programmes in the contemporary globalised world. This means both that India has an important role in the creation and implementation of international space law, and that it can cooperate with other states and organisations as an equal partner. International cooperation has always been a feature of the Indian programme, and is itself a mechanism for developing Indian influence and demonstrating its technological sophistication by working with the leading states in the developed world. India plays a significant role within key bodies such as the UN Committee on the Peaceful Uses of Outer Space, the UN Economic and Social Commission for Asia and the Pacific, the international COSPAS/SARSAT search and rescue system, and the International Global Observing Strategy. Insat 2B launched by the European Space Agency in July 1993 among other things carried search and rescue transponders as part of the international COSPAS/SARSAT satellite system which picks up and rebroadcasts distress signals. India has built SARSAT ground stations at Bangalore and Lucknow, which not only serve the areas adjacent to India, but also give coverage for ten other countries in the Indian Ocean region. India's Department of Space has cooperation agreements with several other countries and space agencies, including Russia, the European Space Agency and NASA. Indeed, reflecting the country's non-aligned status, India has been careful to cooperate with a broad range of states and not focus cooperation with a particular country or group in a way that might compromise its independence.

Although India's overall relationship with the United States has significantly improved since the end of the Cold War, the missile development programme remains a major source of friction between the two countries. India deeply resented the US decision to impose sanctions against ISRO in 1992 after India concluded an agreement with Russia for the purchase of cryogenic engine technology. India went out of her way to stress its determination not to be intimidated by the United States in this regard by pointedly carrying out a test-firing of Agni on 29 May 1992, the date of the first Indo–US naval exercise.[77]

It can be argued that the traditional development/security dichotomy is not particularly enlightening, and that in any case the development of security theory in the past 20 years allows Indian space policy to be more usefully seen as being driven by *security* needs. This is so because the broadened definition of security embraces both the traditional military dimension and extended categories of economic, environmental, societal, and human security.[78]

Prestige and grandeur

Like the Soviet and French space programmes, India's efforts in space research and development have been driven in part by a need to bolster the nation's self-image and international standing. India as a cultural entity and as a nation of people with a common history is millennia older than the present Republic of India, which has been in existence for less than half a century. The modern Indian state is committed to secular, egalitarian values and the pursuit of material prosperity through technological progress. These values are novel in the long history of India, as indeed are the subcontinental boundaries of India itself. At the same time, the modern India, with its rich legacy of a civilisation older than that of Europe, joining a state-system shaped by Europe and pursuing a European political and economic model, has seen a need for symbolism in its foreign policy. India has needed to believe that it has a great deal to offer the family of nations and that it draws on strengths that pre-date the colonial experience.

This feeling grew in the 1950s and 1960s when India saw itself as competing for the moral and political leadership of the Third World with the alternative model offered by communist China. The space programme played an important political role in contributing to India's dominant position in south Asia, in establishing its military and political superiority over Pakistan and in being a credible competitor with China.[79] Above and beyond its role in national development, India's space programme is a symbol of its success as a people and as a state, while demonstrating its ability to compete with the world's great powers in one of the most demanding areas of high technology, and providing, as it does for France, a dramatically *visible* symbol of national achievement. In addition, 'India's space policy is an integral part of the country's foreign policy, which is aimed at strengthening India's role as the dominant power in South Asia and a leader of Third World countries'.[80]

India's programme is a reflection, not only of the pursuit of technological solutions to problems of development, but also of normative considerations, 'facts of technology are a guide, but the contours of a space policy are subject to values, and values pursued in space exploration are related to community expectations'.[81] In this regard, India's space programme could long claim to be one of the most cost-effective in the world, a status it achieved by limiting its ambitions and avoiding purely prestige projects.[82]

At the turn of the twenty-first century, however, the programme acquired a significant new component, with the announcement of long-term goals of manned spaceflight and planetary exploration. These were precisely the goals that the space programme's founder, Vikram Sarabhai, had rejected in favour of a developmental rationale. Sarabhai had insisted in 1968 that India's application of space technology to addressing its development goals 'is not to be confused with embarking on grandiose schemes' and specifically that 'we do not have the fantasy of competing with the economically advanced nations in the exploration of the moon or the planets, or manned space flights'.[83]

Yet in 2006 ISRO proposed starting a human space flight programme, with the first manned flight scheduled for 2014 and aiming to land a man on the

Moon by 2020, before China accomplished the same feat. ISRO's chairman, Gopalan Madhavan felt obliged to defend the switch from Sarabhai's views in announcing the policy shift, declaring that 'That policy – pronounced four decades ago by Vikram Sarabhai, father of India's space programme – had to change for two reasons. We believe that pushing forward human presence in space may become essential for planetary exploration, a goal we have set for ISRO twenty years from now. Secondly, with India's booming economy, costs should not be a hurdle'.[84]

Government approval of the programme was significantly increased by the fact that A.P. Abdul Kalam had become President of India on 25 July 2002. He had worked for ISRO between 1963 and 1982, eventually becoming Director of the SLV-3 programme, before leaving to become Director of the Integrated Guided Missile Development Programme which developed the Agni missile. President Kalam had made known his strong support for a manned programme.

Significantly, it was the launch of an astronaut by China that led India to begin planning for a manned programme of its own, with design work beginning immediately for an adapted GSLV launcher and a two-man space capsule.[85] The shift in Indian priorities suggests that prestige considerations *vis-à-vis* China have become an important factor, and that a space race between the two Asian giants has begun. In addition, ISRO suggested that global competition lay behind its desire for lunar missions, an ISRO spokesman declaring that 'There is now a feeling that, 20 years down the line, other countries would have explored the Moon for minerals and India must not be left behind'.[86]

It is unlikely that the new pursuit of prestige through the space programme will come to dominate it at the expense of the development rationale, but it is clear that India has now added scientific and prestige goals to the development and military rationales that have dominated her space policy for four decades, bringing her programme into line with the rationales typical of the other major space powers.

Chapter 10

China

The long march into space

For many analysts, the Chinese space programme is seen as a classic example of the pursuit of space power, defined as *'the pursuit of national objectives through the medium of space and the use of space capabilities'*.[1] Certainly China's policies in this area have been pursued with consistency over the past 50 years, though China has neither prioritised its space programme over other objectives, nor pursued the full spectrum of uses seen with the United States.

Nonetheless, China, like India, began to construct a programme very early on in the space age. To some extent China was unusual in that, even more than India, there was a pre-existing history of interest in rocketry going back several centuries. China has a longer tradition in rocketry than any other nation, having produced gunpowder-driven rockets for military purposes over a thousand years ago.[2]

In some ways the motivations of China's space programme are similar to those of India, but in others they differ markedly. China, like India, is strongly influenced by prestige considerations. It experienced the brutal ravages of colonialism and imperialism in the nineteenth and twentieth centuries, and is driven in part by a desire to rise beyond the imperialist legacy and to be recognised as a sophisticated and technologically advanced state. Prestige, and contributing to military capabilities that would prevent a return to imperialist exploitation have therefore been central to government support for the missile and space programmes from their inception in the late 1950s.

The Chinese government's focus on prestige arises out of the catastrophes that befell the country in the nineteenth century and the first half of the twentieth. China was invaded, humiliated and partly partitioned between a number of European colonial powers as well as Japan, resulting in the loss of large amounts of territory, the imposition of policies detrimental to China's interests, and massive human rights abuses. The middle decades of the twentieth century were characterised by bloody war with Japan, accompanied by a civil war between the Nationalist and Communist parties, a war from which the Communist Party emerged victorious in 1949. Under the Chinese

Communist Party (CCP) there has been a consistently pursued effort to regain the traditional great power status that was lost in the preceding two centuries. China's perception of itself was a national narrative that perceived the country as 'a great civilisation that had been robbed of its status by well-armed barbarians'.[3] Driven by this dynamic, China's space programme rationale has mirrored those of the early Soviet and American programmes, the desire to 'gain national prestige, and to signal wealth, commitment and technological prowess'.[4]

The original impetus for Chinese rocket developments in the modern era came from military considerations, again echoing the superpower precedent. China's nuclear weapons programme was initiated by chairman Mao Zedong in 1955 and generated a requirement for long-range missiles that could reliably deliver China's nuclear warheads to their targets. The missile development programme was inaugurated in May 1956 when the Ministry of Defence established the Fifth Academy, for missile research.[5]

In 1956 Mao Zedong launched a programme to develop China's scientific base and the space programme has its origins in this development. Following the dramatic launch of Sputnik in 1957, Mao declared that 'we also want to make artificial satellites'.[6] In the years that followed, the Chinese space programme was allowed to develop in a methodical manner, despite the ideologically inspired upheavals that tore Chinese society apart in the late 1950s (the Anti-Rightist Campaign and the Great Leap Forward) and 1960s (the Cultural Revolution). The leadership of the CCP were convinced that the missile and space programmes were of crucial importance to the country, and they were therefore insulated from the turmoil that periodically racked other aspects of government policy.[7] In both the Anti-Rightist Campaign and the Great Leap Forward, the rocket programme was spared the purges and dismissals that affected intellectuals and scientists in other policy areas. Despite the serious economic difficulties being experienced in China by 1960, the government reaffirmed its commitment to heavy investment in space-related programmes.[8]

Initially the Chinese space programme benefited from the support of the Soviet Union. The two communist giants were allied during the first half of the 1950s and the Soviet Union provided considerable technical aid in the development of Chinese nuclear and missile systems. The USSR helped in the creation of China's missile research and development institutes, the two countries signing a bilateral agreement in October 1957, the month in which the USSR launched Sputnik I, and the Soviet Union also provided sample rockets derived from the German V-2 for Chinese engineers to examine. However, in the late 1950s ideological differences between the two countries emerged, and relations deteriorated sharply. As a result the Soviet Union abruptly ended its technical assistance programmes with China. In August 1960 the Soviet Union withdrew all technical support for the Chinese missile development programme. This was a major set-back in many ways, but also acted as a spur to China in developing a self-sufficient technology.[9] Ironically, the Chinese programme also benefited from unintended American support during this period. Two of the leading figures in the subsequent Chinese rocket programme, Chien Wei-Chang and Chien Hsue-Shen, returned from the California Institute of Technology in 1947 and 1955 respectively, as a result of anti-communist pressure brought against them during the McCarthy era in the USA. As the Cold War intensified, Chien Wei-Chang, who had been on the Long March with Mao, rose to become deputy head of the Chinese rocket programme.[10]

Chien had gone to the US as a student in the 1930s under a scholarship programme, held the rank of USAAF Colonel during the Second World War, and went on to become a full professor after the war. He sought American citizenship but with the outbreak of the McCarthy 'Red Scare' was placed under house arrest in 1949, on suspicion of spying for China. In 1955 these pressures led him to return to China. Just over a decade later he presided over the successful first launch of a Chinese nuclear missile in 1966. Even the Soviet Union found the American behaviour astonishing, asking in a 1968 Moscow radio broadcast 'how did this high-ranking US military officer, who was closely connected with nuclear and rocketry research, end up in People's China?'[11]

The Chinese space programme has had strong ties to the military from its inception, when it was placed under the Fifth Academy of the Ministry of Defence. Because of the disruptions threatened by the events of the Cultural Revolution, the space programme was placed under martial law, implemented by the Academy of Space Technology. While those working in the programme were mainly civilians, the authority within the programme was firmly in the hands of the military, who treated the missile programmes as military projects and ensured that the civilian staff working within the programme came under military discipline.[12]

China in the late 1950s felt a strong perception of being threatened by the United States. America had used nuclear weapons to defeat Japan in 1945 and had intervened threateningly in the Korean War (1951–3) and the Taiwan Straits crises (1954–5, 1958). Forward deployed US nuclear-capable forces in Taiwan, Japan, South Korea and the US Pacific fleets posed an unanswerable threat to China, driving an urgent pursuit of nuclear weapon and long-range ballistic missile technology. These pressures only increased in the 1960s after the Sino–Soviet split, when the Soviet Union, formerly an ally, also became perceived as a major threat (and by so doing completed the encirclement of China by the forces of the superpowers). China shared the longest militarised border in the world with the Soviet Union. The USSR began deploying nuclear weapons against China in 1966, and there was heavy conventional fighting between Chinese and Soviet forces on the borders in 1967 and 1969. Much of the Sino–Soviet border region was thinly populated and made up of difficult terrain, so that the most efficient and least provocative possibility for monitoring Soviet military activities, particularly close to the border, was the use of reconnaissance satellites.

Despite these geopolitical tensions, China also began a civilian space programme at this time, initiating a programme for launching sounding rockets in 1960,[13] but the key Chinese space launchers have been derived from modified long-range ballistic missiles, rather than from developments arising out of the civilian-sounding rocket programme, and there has therefore been a strong historical linkage between the pace of nuclear weapon and ballistic missile development, and the advance of space launcher technology in the Chinese programme. The first Chinese nuclear weapon test took place in 1964, and in 1966 the first missile-launched nuclear explosion took place. The warhead was carried on a CSS-1 missile. Later development and the addition of further missile stages produced the CSS-3 ICBM, of which the subsequent Long-March-1 space launcher was a modified version.

In the late 1960s China was again convulsed by political turmoil, this time the Cultural Revolution, and once again the space programme proved one of the few policy areas spared the worst effects. The space programme 'found itself in the enviable position of receiving strong support no matter which of the contending forces held power'.[14] Prestige and military considerations were central to this, because of the strong nationalism that was central to all the rival factions. Mao Zedong was strongly supportive of the missile and space programmes, because these technologies were seen as crucial indicators of power and capability in the dangerous confrontations with the two superpowers.[15] Foreign Minister Zhou En-Lai was also a strong supporter of the programme, because he believed it would be crucial in rebuilding China's lost prestige in the international community. Chien Hsue Shen, the director of the Chinese space programme, deliberately cultivated the patronage of Zhou En-Lai, and this was crucial to the programme's survival during subsequent periods of domestic political turmoil.[16] Even Mao's opponents supported the programme, because they saw it as reflecting prestige and a reputation for competence on themselves.[17] Insulation from the political excesses of the Cultural Revolution was also achieved by the imposition of martial law and the placement of the space programme under the control of the People's Liberation Army. A new Space Academy responsible for space research and development was created and placed under the authority of the Committee of Science and Technology for National Defence, part of the Ministry of Defence. By the time the Cultural Revolution came to an end, China had launched a total of seven satellites, including its first recoverable reconnaissance satellite.

This prioritisation of technological progress over political correctness marked the beginning of the 'techno-nationalism' that would become a key distinguishing feature of Chinese policy by the end of the century. Even at the height of the Cultural Revolution, 'technology and Western military concepts had begun to displace politics and ideology as the underpinnings of China's military policies'.[18] Along with more flexible and marketised economic policies, the commitment to rapid technological development would mark out the subsequent Chinese political programme. Countries in this phase of their development, which have the potential to increase significantly their economic performance and international influence, 'tend to take an especially nationalistic approach to technological development'.[19] The historical experience of the Celestial Empire at the hands of the 'well-armed' barbarians in the nineteenth and early twentieth centuries had already made technological progress a fundamental objective of China under Mao, as can be seen in the policies promoted from the mid-1950s. This conviction only strengthened from the mid-1970s as China emerged as one of a number of governments who saw high technology, and particularly the aerospace and nuclear industries, as the key both to their economic development and the recapture of the international position and status that they felt was their national birthright.

The first successful launch of a satellite by China occurred in April 1970. This was the launch of the DFH-1 (Dong Fang Hong, or 'The East is Red'). The satellite broadcast a stirring revolutionary song of the same name for the duration of its 26 days operating in orbit. China lauded the feat as a victory for the Ninth Party Congress of the CCP, and evidence that the Party was 'achieving greater, faster, better and more economical results in building socialism, and by grasping revolution, promoting production and other work, and preparedness against war with concrete action'.[20] China's foreign minister, Zhou En-Lai, insisted that the post-launch communiqué include the words, 'we did this through our own unaided efforts'.[21] The hyperbole notwithstanding, this was a dramatic technological achievement. China became only the fifth country in the world to launch a satellite into orbit on one of its own launchers. The launch of the recoverable satellite in November 1975 meant that China became only the third country to master this technology, the others being the two superpowers.

Ironically, after the death of Mao it was the gradual emergence of a new Chinese ideology, capitalism, which subsequently placed the programme under great pressure. In an environment of economic weakness, the Chinese government emphasised the imperative of economic development, and the requirements of what has been described as 'Market-Leninism', meant that the space programme had to justify its existence and its budgets in a new way. China's new leader Deng Xiao-Ping insisted that China's priority should be economic development. In a key speech in 1978, he insisted that the space programme must orient itself towards helping to achieve China's wider social and economic goals.[22]

With a new emphasis on the commercial aspects of space development, the Chinese programme shifted its focus from overwhelmingly military considerations towards income-generating applications, such as communications satellites and launch services. China was aided in this move by the fact that the United States suffered the *Challenger* disaster in 1986, and the European *Ariane* launcher also had problems in this period, and this made Chinese launch services an attractive alternative for commercial customers. In May 1985 China had announced that it was making its launch capacity available for satellite launches for other countries and corporations. To market its launcher and satellite services it established the China Great Wall Industry Corporation. The name chosen for the corporation is significant. As with other elements of the programme such as the *Long March* rocket, the naming of elements of the space programme is used to establish mental linkages with heroic or impressive elements of China's past and reflects the centrality of national recovery and prestige as drivers of the space programme.

Aided by highly competitive prices and the difficulties being experienced by other launch providers, China began to acquire a significant share of the global market for satellite launches during the 1990s.[23] In the late 1990s China's market share peaked at over 10 per cent, though it fell back subsequently.[24]

From the mid-1970s to the mid-1980s the Chinese space programme struggled to adapt to the new environment in which it found itself, and for the first time since its inception it lacked strong support from the CCP and government, and suffered a consequent drop in prestige. But from the mid-1980s onwards the space programme began to regain its former prominence. It benefited from a number of parallel developments. In the first place, it had a clearer sense of the missions that the government wanted it to perform. Second, morale and self-confidence within the programme revived as it became clear that it was once again receiving strong support from the highest echelons of the CCP, particularly from Premier Zhao Ziyang. In 1986 Zhao Ziyang convinced the rest of the Chinese leadership to make the space programme the priority technological programme in China. The following year Deng Xiao-Ping designated Zhang as his successor, thereby assuring the space programme of long-term support.[25] Finally, by this period China had accumulated a critical mass of scientific and technological expertise and experience. Thus by the late 1980s the space programme was acknowledged to lie at the heart of China's efforts to develop a strong scientific and technological infrastructure as the basis of its future development.

Whereas at the start of the decade China had been launching one satellite a year, by 1989 it was able to launch 24 satellites in a single year, of which 11 were recoverable. The satellites were placed in a variety of different orbits and performed a range of different functions. These included meteorology, telecommunications, remote sensing of natural resources, maritime navigation, materials processing and biotechnology.

As noted earlier, one possible conceptual framework for interpreting the socio-political role of the Chinese space programme is that of 'techno-nationalism'. It is clear that China sees technology as the key to its long-term economic development, and values the space programme's potential for generating advances in cutting-edge technology. For a number of analysts, 'techno-nationalism is the twenty-first century equivalent of the earlier developmental nationalism that had stemmed from colonial subjugation'.[26]

In this view, whereas the key variable during the Cold War superpower confrontation was political alignment, which influenced both the relative strength and influence of the superpowers and the political and economic development of the allied or associated states, in the contemporary international system the development of advanced technology has now become the key system variable in the way that alignment previously was, and geotechnological manoeuvring has supplanted geopolitical competition.[27]

This is a useful way of thinking about the role of the space programmes for rising great powers with continuing development problems to overcome, such as China and India. In such countries the pursuit of expensive space programmes seems a problematic political choice in the face of the continuing problems posed by the needs of very large and deeply impoverished rural populations. This is particularly the case as such countries move into the hugely expensive and prestige-driven manned spaceflight phase of the programme, as both China and India are now doing. Techno-nationalism therefore provides a 'useful framework for understanding the motivations of developing great powers such as China'.[28] In 1986 the Chinese government designated space as China's highest priority technological programme.[29]

Relations with the United States

The United States has been markedly negative in its attitudes towards the Chinese space programme compared to other key actors such as Europe and Russia, which have taken a more positive and cooperative stance towards China. For example, China has not been invited to participate in the International Space Station programme. Annual Pentagon reports on PRC military power suggest that the United States sees China as the country most likely to challenge its dominance in space. The Department of Defense's 2004 and 2005 annual reports on China's military power identified China's *counterspace* developments and noted that it was working on plans to field anti-satellite weapons.[30]

During the 1980s China negotiated a number of contracts with American companies for launching US satellites on the Chinese CZ-3 launcher. However, concerns from the US government about unfair competition led to China signing a Memorandum of Understanding with the US, in which it agreed to launch no more than nine American satellites over the subsequent five years. Even so, following the Tienanmen Square incident in June 1989, the administration of President George H. W. Bush decided to impose a range of sanctions against China. One of these was a decision only to authorise export licences for American satellites on a case-by-case basis. Nevertheless, the American Asiasat-1 was launched in April 1990, China's first international commercial satellite launch.[31] The export ban was subsequently lifted in April 1991.

Sino–US space relations became problematic again in the mid-1990s, following American allegations of Chinese technology espionage and the deliberate transfer of sensitive technology to China by US corporations such as Hughes and Lockheed Martin. A significant lobby had emerged within the US Congress who were ideologically opposed to communist China, and highly suspicious of China's military modernisation programme. In the face of such political hostility it became effectively impossible for China to gain export licences for US satellites, or even to launch Chinese satellites with significant American-made components. In response to this situation China, at considerable cost, replaced American-made components from its own systems and redirected its commercial launch strategy towards markets in Europe and Asia.[32]

China in turn has no reason to accept America's self-appointed hegemonic dominance of space. The United States has made no secret of its intention to exercise effective 'space control', providing dominance in peacetime and hegemonic monopoly in wartime. Moreover, Chinese observers noted with alarm the 2001 US military space exercise, in which China was the designated enemy. Senior Chinese aerospace officials have therefore argued that China must 'develop advanced weapons for space warfare'.[33]

In any conflict between the United States and China, the latter would be operating with hugely inferior military technological capabilities. The rapid and overwhelming American victory over Iraq in 1991 demonstrated to Chinese leaders with shocking clarity the enormous military advantage US space dominance and associated information superiority provided for American forces. Chinese strategists have therefore looked for potential American vulnerabilities where a strategy of asymmetric warfare might be brought into play. One area where the US is clearly asymmetrically vulnerable in this way is in its reliance on military space assets as a force multiplier. Chinese analysts have speculated that 'for countries that can never win a war with the United States by using the method of tanks and planes, attacking the US space system may be an irresistible and most tempting choice'.[34] It would be surprising therefore if China was not researching ways to exploit this situation. The United States has itself publicly declared its alarming vulnerability in this area. The Rumsfeld Commission report of 2001 warned of the rapidly increasing US dependence on military space 'and the vulnerabilities it creates'.[35] The previous year, the Director of the Defense Intelligence Agency had testified to the Senate Intelligence Committee that China was actively engaged in development programmes designed to capitalise on this American vulnerability by deploying weapons and other anti-satellite systems that would be able to negate America's military satellite systems in wartime.[36]

Certainly US 'space control' doctrine would be challenged by China's efforts to deploy both passive military space systems and her research programmes into various ASAT technologies. These would mean that the US could not rely on diplomacy to protect its satellites, as it did in the Iraq wars of 1991 and 2003. The more sophisticated Chinese offensive and defensive counterspace capabilities proved to be, the higher up the ladder of escalation the US would be forced to ascend in order to overcome them.

China has adamantly opposed the American ballistic missile defence programme and its encouragement of regional allies to participate in the development of ballistic missile defence technology. This is hardly surprising given the limited numbers of strategic nuclear weapons that China has deployed as part of its deterrence posture. China has pursued a 'minimum deterrent' strategy, and having a comparatively small nuclear force means that its capability would be made vulnerable with the deployment of even a limited defensive system. While it is possible for China to develop countermeasures to any defensive system, this creates additional and unwanted technological and financial demands for China. In this regard, China has benefited from the 'strategic partnership' formed with Russia in the mid-1990s which has given her access to advanced data and technology acquired by the Soviet Union during its Cold War confrontation with the United States.

Even though the United States claims to wish to deploy only a minimal capability suitable for intercepting individual launches from 'rogue' states, this hardly reassures China given that even a limited system would threaten their limited offence. Moreover, they are well aware that the US system could be upgraded to a more capable system if a later administration wished it, and indeed the National Missile Defense programme specifically provides for this eventuality. China is also aware that the United States has demonstrated in the past that it does not see itself as constrained by treaties and agreements it has signed, if it decides that these no longer serve American interests. China perceives the Bush administration's deployment of the ground-based Midcourse Missile Defence system as a clear and decisive first step on the road to the weaponisation of space. Chinese military specialists have argued that the missile systems deployed on America's west coast could also be used as ASAT systems.[37]

Chinese concerns were increased after 2000 by a sense that US support for the weaponisation of space was gaining momentum with the advent of the George W. Bush administration. Ominous for China were the demands in the Rumsfeld Commission Report that the United States move quickly 'to ensure that the president will have the option to deploy weapons in space',[38] and the American withdrawal from the Anti-Ballistic Missile Treaty in 2002. China also took note of the language used in the 2004 Counterspace Doctrine paper,[39] and the 2006 revision of US National Space policy.[40]

China is also opposed to the weaponisation of space in a more general sense, because it sees space as the new 'high ground', control of which provides enormous military advantages in relation to terrestrial conflict. Chinese military analysts see the weaponisation of space as a uniquely threatening development given China's current inability to respond effectively to the threat it represents, a position they have held since the period of the US Strategic Defense Initiative in the 1980s.[41] In any case China's overall policy goal is national economic development, in which the space programme plays an important role. As such, China's preference is for space to be maintained as a weapons-free sanctuary supportive of its overall development policies. Nevertheless, on 11 January 2007, China fired a ground-launched missile into space, destroying one of its older satellites at an altitude of 530 miles. The successful test demonstrated that China would be able to target American satellites during a conflict, and would inevitably raise the tensions in any future crisis.[42] It also suggested that China is becoming more pessimistic about the chances of achieving a space arms control regime, since its test would significantly complicate efforts to produce such an accord.

Rather than directly confront the United States' space hegemony, China has sought to negate it through a policy of encouraging multipolar modifications to the international space regime, which would bring a 'cooperative balance into play in space'.[43] For example, China has sought to strengthen the international legal regime relating to space, particularly through the United Nations. The 2000 China White Paper on Space Policy stressed the importance of the United Nations to Chinese thinking about the international space regime. Efforts in this regard focus on 'controlling orbital debris, space traffic and enhancing transparency by providing details on satellites launched'.[44]

China also continues to pursue the possibility of participation in the International Space Station. Having indigenously developed its own manned space programme China is in a stronger comparative bargaining position than it previously was. In addition, since Chinese manned space systems are based on Russian designs they 'can easily be made compatible and interoperable with the ISS, which relies on many Russian components'.[45] However, it has made it clear that if this does not materialise, then it will go ahead and develop a second international space station in partnership with other countries, in order to encourage the development of multipolarity in space. This would allow other states to use space in a way which reduces the American dominance in a way that no state could hope to do purely through its own efforts. It would also allow other states, particularly those from the developing world, to play a genuine role in shaping future international space developments from the outset, rather than simply participating in an environment shaped by others.[46]

The relationship with Russia has become a crucial component of Chinese space progress since the end of the Cold War. In 2006 Russia announced that it was about to conclude a lunar exploration agreement with China under which the two countries would carry out joint projects on the Moon and on

Phobos, the larger of the two moons of Mars. The Russian–Chinese Space Exploration Commission was represented at Prime Ministerial level on the Russian side, a clear indication of the importance attached to Sino–Russian space cooperation by the Russian government.[47]

Perhaps recognising the risks involved in deliberately excluding China from American-led space activities, in April 2006 US President George W. Bush suggested to Chinese President Hu Jintao that NASA might pursue significant cooperation with the Chinese space programme in future years. A specialist at the National Security Council suggested that the President believed that such cooperation made strategic sense since the United States and China were 'the two nations on the earth with the most ambitious space programmes in the twenty-first century'.[48]

China is also developing a close working relationship with Europe (both ESA and the EU) in space-related matters. The most dramatic example of this is the Galileo satellite system, in which China has a 5 per cent stake. China has also collaborated with the European Space Agency in developing the small *Doublestar* satellites and has received Earth-observation data from ESA satellites through the *Dragon* programme.[49]

Nevertheless, the military rationale has remained central to China's own programme. Under Deng Xiao-Ping's 'four modernisations', China's military/space community focussed on the development of reliable ICBM and SLBM technology to provide an effective nuclear deterrent, and on the development of communication satellites.[50] These satellites were prioritised over other applications satellites because as a very large and mountainous country with large desert areas, China needed reliable long-range military communications for command and control, and also the ability to use its satellites to target its long-range nuclear weapons.

The general pattern of China's military space usage is very similar to that of Russia and the United States, with particular emphasis on the deployment of navigation and communication satellites. However, its overall capability does not match those of the former superpowers, because it has far fewer satellites in orbit at any one time than they do, with most of the operational satellites being used for communications, rather than the sensor platforms that are typical of the United States.[51]

Though China has launched a number of military communication satellites, the majority of communication satellites orbited have been for civilian purposes. Communications satellites have become a priority element within the civilian space programme, because they are seen as a major benefit in terms of achieving national economic development goals. Chinese analysts have stressed their value 'for education, government, transport and the financial and commercial sectors of the economy'.[52] While the military communications satellites are less capable than their western counterparts, China is far less dependent on satellites for such communications, and has maintained and upgraded terrestrial communication systems. This means that, in the medium term at least, 'China's lesser dependence on space communication systems also means they would be less impacted by an adversary's hostile space activities'.[53]

China was slower to develop applications satellites than it was launchers, the first civilian communications satellite not being launched until 1986. Chinese satellites are less capable than western equivalents, and China supplements its own capabilities by leasing capabilities from other satellite providers. China has also worked cooperatively with western companies in the development of its latest generation satellite technology, for example with Messerschmitt–Beolkow–Blohm in the development of the DFH-3 satellite.[54]

Cooperative international commercial ventures have fewer security risks attached than the development of defence-related technology, but they may still be politically problematic from a Chinese perspective. In October 1993 'China banned private reception of satellite TV, despite the fact that Chinese state-run corporations had significant interests in both AsiaSat and APT companies which relayed the broadcasts to China'.[55]

Earth-imaging satellites are the second priority for China after communications satellites. Apart from their crucial role in contributing to military security, they are seen as crucial for 'monitoring of natural disasters, followed by developing natural resources, particularly agriculture'.[56] China is a mountainous country with large deserts, making it a major task to feed its population of more than one billion on the arable land available.

China's military space programme has been curiously slow-paced and limited since its inception. Despite the obvious advantages of reconnaissance satellites to the PLA for example, China waited until the end of the twentieth century before it began using electro-optical imaging satellites rather than the ancient film-recovery systems that it had relied upon for three decades. Thus it is only very recently that China has acquired a real-time satellite imaging capability.

China launched its first navigation satellite in 2000. The system was completed with a second satellite launched the same year. Although it is a very limited system with only two satellites, both in geosynchronous orbit, it does provide China with 24-hour coverage over the Chinese mainland and adjacent waters. Like other countries which have benefited from providing commercial space services, China also fears the danger posed to their space industry by military space threats, so that developments in the civilian and military space realms are inextricably linked in Chinese eyes.

For most of its history, the Chinese space programme has not emphasised exploration for its own sake, or a manned programme. These were seen as low priorities because they did not make a direct contribution to defence or development.[57] An early attempt to develop a manned programme in the 1970s was abandoned because of the costs involved and the limitations of China's space technology in that period.[58] Planning for a Chinese manned spaceflight programme began as early as 1966, and in March 1971 China became only the third country in the world to select a squad of astronauts as part of project Shuguang (Dawn). However, in the face of domestic political infighting and competing resource claims, political support for the programme could not be sustained.[59]

The manned programme, 'Project 921', officially began in 1992. The Shenzou (divine vehicle) space craft was developed for launch on the Long-March 2F rocket. The first four Shenzou launches were unmanned, 'but the fifth and sixth carried *yuhangyuan*, 'travellers of the Universe', to use the Chinese term for astronauts'.[60] The Shenzou was recognisably similar in design to the Russian Soyuz spacecraft, and provides clear evidence that China had clearly benefited from the growing space cooperation that characterised Sino–Russian relations in the 1990s. However, it was not simply a Chinese copy of the Russian design. China modified the original design significantly. Shenzou is larger than Soyuz, and potentially capable of carrying as many as four astronauts into orbit. Significantly, the latest generation of Chinese space launchers, the Long March 5 series, will have a lifting capacity capable of orbiting a space station of the Salyut/Mir class.[61] Given that Russian technology and information transfer have been so significant since the end of the Cold War, and that China has spoken openly of its plans to orbit such a space station, it is likely that Russia has given China access to information about the design of the Salyut/Mir space stations and therefore a Chinese space station launch before 2010 is quite possible. Should China do so, it would provide her with the ability to use the station for political purposes via joint missions with other countries, just as the Soviet Union did with the *Intercosmos* programme in the 1970s and 1980s.

The manned programme represents an interesting return to prestige considerations for the Chinese space programme. Prestige was certainly an important driver at the beginning of the programme in the late 1950s and throughout the 1960s, but was then eclipsed by more mundane considerations, particularly defence and economic criteria. But with China having 'arrived' as an important international player in the past ten years, its government is seeking both to consolidate that position in meaningful ways, such as membership of the WTO, and to acquire the trappings of great power status, such as a blue water fleet and a highly visible space programme. Manned spaceflight remains, as it always has been, the most dramatic symbol of a vigorous and technologically leading edge space programme.

The recreation of a manned space programme not only necessitated a major investment in the additional infrastructure required for manned flight, it also required a considerable diplomatic offensive. The Chinese astronaut squad went to Russia for their specialist training, China had to establish a network of tracking stations overseas, in Tarawa, Namibia and Pakistan, and had to construct and deploy a fleet of tracking ships for 'blue water' deployment during manned missions.[62]

As a developing country with significant budgetary constraints, however, China has been historically cautious in its approach to deep space exploration and manned space flight. Former leader Deng Xiao-Ping cautioned the leaders of China's space programme not to focus on expensive missions with a low pay-off for China's development plans, such as flights to the Moon.[63] It is China's growing wealth, coupled with the growing centrality of prestige considerations that have changed this perspective. Manned space flight is far more expensive and technologically demanding than is unmanned space exploration and is ultimately pursued by any state for political reasons. The economic and military benefits of a manned programme are minimal, it is the political benefits alone that make it attractive to Chinese policy makers.[64]

In this regard, it is important to note that the prestige dimension of the Chinese space programme is directed as much at domestic as at international perceptions. China's phenomenal economic growth in the past two decades has not been without social and political costs, and has seen growing problems of corruption and social injustice. Even official Chinese government figures admitted to 74,000 demonstrations against local or national government policies in 2004.[65]

Critics of the space programme see it as an unnecessary diversion of resources away from China's economic development efforts, particularly in the context of the October 2004 decision by the Central Committee of the CCP to 'enhance efforts in adjusting income distribution and alleviating the widening wealth gap'.[66] For China's government therefore, there are pressures similar to those facing India. Like India, the Chinese government has responded to such criticisms by emphasising the space programme's contribution to national development goals.[67] External validation of such goals is also emphasised. Thus, for example, when Spanish space officials visited China in 2003, the subsequent press release on the discussions noted that 'the two parties agreed unanimously that space technology plays a crucial role in promoting national economic development and people's living standards'.[68]

The President of the Chinese Academy of Space Technology stressed in 1997 that Chinese satellites had 'played a very important role in developing the national economy, promoting scientific and technological progress and developing culture and education'.[69]

Achievements claimed for the programme included the use of satellite data for land survey, geology, water conservation, oil prospecting, mapping, environmental monitoring, earthquake prediction, railway line selection, and archaeological research.[70]

China's diplomacy is sensitive to the external perceptions of the space programme, particularly in relation to its military dimension. In 2002 China and Russia jointly submitted a working paper to the UN Conference on Disarmament proposing a ban on the placing of weapons in space. Chinese space officials are keen to downplay any suggestion of an emerging military space competition involving China. Such an effort, it is argued, would be economically ruinous for China, and therefore while the dynamism of the Chinese programme inevitably conjures up images of a new space race, 'this is not a competition like the Cold War'.[71] Nevertheless it may still carry echoes of earlier competitions. It has been suggested, for example, that the growing cooperation between Europe and China, particularly in regard to space technology, is driven not only by a desire to gain mutual economic benefits, but to some degree by a common wish to balance against the preponderant power of the United States.[72]

China has taken symbolic steps to project a more reassuring face for its programme. During the 1990s the administrative control of the space programme was taken away from the PLA and given to a new government agency, the China National Space Administration, reporting to the Science, Technology and Industry Commission for National Defence. The programme remained central to the defence effort, and was clearly still within the overall control of the armed forces, but at one step removed.

Like other dimensions of Chinese policy, the space programme reflects the changes in overall foreign policy goals and approaches that have occurred in the past two decades. China has emerged from its Maoist isolation to become a responsible actor within the international community. It is eager to be perceived as such and to be seen operating in 'accord with international standards, particularly in economic and scientific practices'.[73]

From the perspective of the international community there is everything to gain from encouraging the increasing integration of Chinese space activities with those of other states. In terms of the manned programme for example, the loss of the space shuttle Columbia and subsequent reliance of the US on Russia to maintain the servicing of the ISS 'has painfully underscored the need to have redundant capabilities for launching humans into space'.[74] The emergence of China in this regard is of great significance, particularly given that the Shenzou can carry a larger complement than can the Russian Soyuz.

Conclusions

For China, prestige considerations are central to the space programme, helping China to finally shake off the memory and image of the humiliated prostrate China of the nineteenth and early twentieth centuries. It represents the rebirth of China as 'the Celestial Kingdom' this time in a practical as well as a figurative sense, as its emergence as a genuine space power cements its great power status in the post-modern age. The original Long March of Mao and his army was an epic struggle that formed the centrepiece of Chinese communist propaganda for decades afterwards. The manned space programme also 'will be couched in a language of heroes and glorious deeds' and project a national image of vision, capability and commitment.[75] The Long March to Space and the rebirth of the Celestial Kingdom are fitting metaphors for China's path to a global leadership role.

Chapter 11

Cooperation and competition in the post-Cold War era

In the post-Cold War period a number of themes have been prominent, but three stand out. The move towards the increasing use of space for military purposes by all space powers, with the US pursuit of 'space control' has already been discussed. The other two key themes are the growing internationalisation of space activity, particularly with the International Space Station and the dynamic influence of the Chinese space programme on the programmes of other countries.

The international space station

In many ways, space exploration is a naturally 'federative' activity which encourages international cooperation on a variety of grounds. Nevertheless, the experience of such cooperation has revealed that it can be extremely problematic to make work effectively in practice, and on occasion may be a cause for tension and recriminations between governments rather than mutual congratulation. There are significant difficulties involved in the establishment and implementation of long-term joint space projects, 'during which any number of economic and national policy imperatives may intervene to disrupt schedules and commitments'.[1]

One of the features of the American space programme has been that it has often lacked a clear sense of direction. This may seem an odd thing to say about the country which placed the first humans on the Moon and developed the space shuttle. Historically however, the US programme has developed in spasmodic surges rather than through consistent evolution. The programmes of Europe, Russia, China and India have been less spectacular in most ways, but have exhibited a certain coherence and evolutionary development over time as they sought to meet a set of reasonably clear objectives within severely constraining financial realities.

The American programme in contrast has often lacked a clear sense of purpose, outside of the military rationales. American leaders do not seem to have had any clear idea of what they expected from the US programme, other than military security and an arena for American companies to make profits. In the 1960s the US had a clear purpose, but it was an essentially negative one, to beat the Soviet Union to the Moon and reclaim the technological and political leadership that had been threatened by the Soviet space successes from Sputnik onwards. Once that goal had been achieved, the administrations of Nixon and Carter demonstrated no clear vision as to what they expected from the space programme. According to retired Apollo astronaut and the last person to walk on the Moon, Eugene Cernan, 'we really haven't had a space programme since we came back from the Moon. There have been a series of space events, but there was always no agenda, they never led anywhere'.[2]

Under President Reagan however, there was an attempt to re-energise the programme with clear political goals. The most dramatic and eye-catching was the Strategic Defense Initiative, a focused attempt to use space technology to reverse the strategic nuclear determinism that had dominated US security since the early Cold War period. Less noticed at the time was a second politically inspired initiative, designed to provide a focus for NASA's manned space activities, as Apollo had done in the 1960s, and once again to use it to emphasise American technological and political leadership of the free world. This was the space station programme.

NASA had orbited a space station in the early 1970s, sending three crews to the *Skylab* station, which had been built from modified Apollo hardware after the final lunar missions were cancelled for budgetary reasons. Originally, it had been hoped that the space shuttle would come into service in the late 1970s and be able to take further crews to Skylab. However, the shuttle was delayed and Skylab burnt up on re-entry to the atmosphere in 1979. The only Americans to fly in space between 1973 and 1981 were the astronauts on the 1975 Apollo–Soyuz rendezvous mission.

During the 1970s, however, the United States had three successive Presidents who had only lukewarm enthusiasm for the space programme and the manned programme in particular. In 1978 the Carter administration had issued PD-42, a presidential policy document outlining US space policy goals, which specifically stated that 'it is neither feasible nor necessary at this time to commit the US to a high-challenge, highly-visible space engineering initiative comparable to Apollo'.[3] Nevertheless, Carter's Under-Secretary for the Air Force testified to Congress in 1980 that 'the next major commitment after the Space Shuttle is completed should be the development of a permanently orbiting manned space station'.[4] President Reagan, in contrast to his predecessors, was an enthusiast for the space programme, and a politician deeply suspicious of the Soviet space effort. In a speech as Governor of California in 1971, he had suggested that the orbiting of the first Soviet space station was an ominous development given that 'those who control space may hold an unbeatable military advantage'.[5] His choice as NASA Director in 1981 was James M. Beggs, a strong advocate of an American space station. During his confirmation hearing, he was questioned by Senator Harrison Schmidt, himself a former astronaut. Schmidt suggested that the space programme had been drifting without a clear goal since it had met President Kennedy's lunar landing challenge, to which Beggs replied that 'the next logical step is a space station'.[6]

Many of the members of the Reagan administration were far less enthusiastic about the project. The President's science adviser, budget director and Secretary of Defense were all strong opponents, with Defense Secretary Weinberger noting that NASA could provide no military justification for the proposal.[7] While Weinberger objected that it would divert space funding away from the defence-relevant SDI programme, science adviser Keyworth's argument was that NASA should be pursuing a more ambitious goal, such as a manned lunar base or a mission to Mars.[8]

In making the case for the station, Beggs focused heavily on political benefits. Beggs argued that the space station would be an important operating base, that it would counter Soviet, European and Japanese challenges to American space leadership, that both the European Space Agency and the Japanese space agency would be interested in participation and could contribute a significant portion of the necessary funding, and that the space station would have major foreign policy advantages for the United States.[9] The West German government was strongly supportive, due to feeling pressured by domestic public opinion and the Warsaw pact states over its decision to allow the deployment of US cruise and Pershing nuclear missiles on German territory. The space station was seen as a civilian project that would project a more positive image about the collaborative efforts of the western allies.[10]

In his 1984 State of the Union Address to Congress, President Reagan announced that 'We can follow our dreams to distant stars, living and working in space for peaceful, economic and scientific gain. Tonight, I am directing NASA to develop a permanently manned space station and to do it within a decade'.[11] Both the phraseology used, and the deliberate choice of a ten-year deadline were deliberate echoes of President Kennedy's commitment to land Americans on the Moon before the end of the 1960s. Unlike the earlier call, however, Reagan's was for an American-led effort, rather than simply a US commitment, and like the Strategic Defense Initiative therefore, there was a deliberate attempt to bring allied and friendly countries under the umbrella of the American programme. Reagan declared that 'NASA will invite other countries to participate so we can strengthen peace, build prosperity, and expand freedom for all who share our goals'. The President had already consulted key foreign leaders on the proposal prior to the speech, and the State department favoured their inclusion, since it would 'demonstrate the strength of American power by allied cooperation rather than unilateral action'.[12]

The terms Reagan used, associating the project with 'peace', 'freedom' and 'prosperity' were both a restatement of what he considered quintessential

American values, and an attack on the totalitarian values and policies of the Soviet Union. This political symbolism would be reinforced in July 1988 when Reagan designated the project as *Space Station Freedom*. The space station was therefore very much a Cold War symbol.[13] Soviet leader Mikhail Gorbachev responded in kind. The follow-on Soviet space station, scheduled for launch in early 1986 and due to be called Salyut 8, was renamed *Mir* (meaning 'Peace'), on his orders. Unlike its predecessors, which had two docking ports, *Mir* had six, and was clearly designed to be the hub of a much larger orbiting complex.[14]

For President Reagan, the SDI and the space station were both parts of a larger overall Cold War framework of using technological programmes to develop US technological advantage, bind America's allies closer, and to push the struggling Soviet economy to its limits. The space station is seen by some analysts as 'the civilian equivalent of the strategic defence initiative proposed by the Reagan administration during this critical period of the Cold War'.[15] With a second Cold War underway, Reagan was trying to bankrupt the USSR, not only by re-igniting the arms race with SDI, but by re-launching the space race through the space station.

However, by this time the Cold War was fading dramatically and would be over completely within three years, and the American political establishment did not react to Reagan's clarion call in the way they had to John Kennedy's. By 1987 Reagan himself had abandoned his Cold War rhetoric and developed an effective working relationship with his Soviet counterpart. One of the fruits of the improving US–Soviet relations was the 'Agreement concerning cooperation in the exploration of outer space for peaceful purposes', signed by the foreign ministers of the two countries in April 1987.[16] Congress authorized only $200 million for the initial work, rather than the $967 million the administration had asked for to allow NASA to begin work on the station. Congress' continuing scepticism about the space station project, and its refusal to fund it at the levels NASA felt necessary led to the resignation of the NASA Director James Fletcher in 1989. The budgetary shortfall meant that in 1989 NASA began scaling down its proposed design in an effort to remain within the financial constraints imposed on the project.

The space station project was also badly affected by the loss of the space shuttle *Challenger* in an explosion soon after lift-off in January 1986. It was not just because of the loss of the shuttle and the associated years of delay before shuttles were deemed safe to fly once more. These delays set back the planned construction schedule for the station, since the shuttle was needed to fly the components into orbit. In many ways even more damaging was the report of the Rogers Commission into the Challenger disaster, which catalogued a huge series of errors by NASA, and strongly criticised the poor safety culture at NASA and the Agency's managerial failings. The commission's report had the effect of making congressional legislators more suspicious of NASA and less willing to fund its ambitious proposals.

The scaling down of the station plans by NASA produced extreme dissatisfaction from the international allies that the project had attracted. The partner countries felt that the United States was failing to consult them on design changes that had huge implications for the elements that they were contributing to the station. The European Space Agency was sufficiently alarmed to announce that it would withdraw from the project if the continuing modifications meant that their *Columbus* orbiting module would not be launched before 2000.[17]

Even so, it was to a large extent because of the international implications that Congress did not terminate the project entirely. In 1991 the House of Representatives overturned a recommendation by one of its subcommittees to cease funding the station, because of the domestic political costs and the likely international repercussions of such a decision. After travelling to discuss participation with the allied governments, NASA Director Beggs had reported to Secretary of State Schultz that 'our principal Allies are moving quickly, or have already moved, to take political decisions to participate. And their reactions clearly showed appreciation for the major foreign policy benefits that will flow from open and collaborative cooperation on such a bold, visible and imaginative project'.[18]

In reality the allies were responding more coolly than Beggs suggested. They recognised that participation in the space station would tie up a large proportion of their space exploration budgets, and were not inclined to take such a momentous decision lightly. At the same time they recognised that it would be difficult to reject the American offer without 'compelling reasons' for doing so.[19] Nor did America's allies react in a uniform manner. While the Europeans were cautious, the Japanese government was enthusiastic, welcoming the chance to cooperate with an 'anticommunist bloc in the field of basic science technology'.[20]

Within the United States there were concerns that the international collaboration would transfer control from NASA and the United States to foreign agencies. Similar concerns afflicted the allies, who worried that they would have insufficient influence on the project and that their industries would not receive a reasonable share of the contracts.[21] Nevertheless, the Intergovernmental Agreement (IGA) between the United States and its international partners was signed in September 1988.

The international space station project was very much a Cold War creation, a tool in the superpower confrontation that had escalated once more in the early 1980s. For the international partners, participation was both a way to energise their own space programmes through collaboration with the highly experienced NASA, and a way of enabling them to demonstrate their credentials as firm allies of the United States in the renewed confrontation with the Soviet Union. The United States remained the hegemonic partner, in the space station as much as in NATO. Article 7.2 of the IGA affirmed that the United States would 'be responsible for overall program management and coordination of the Space Station'.

The winding down of the Cold War at the end of the 1980s removed the rationale of superpower competition that had underpinned political support for NASA and the American space programme particularly during the 1960s. It was noticeable that when President George H.W. Bush unveiled his 'Space Exploration Initiative' with a speech in 1989 it was anti-climatic. The President promised a new policy involving 'manned exploration of the Solar System', and the following year echoed President Kennedy by declaring that the United States should commit itself to the goal of ensuring that by 2019 the American flag would be planted on Mars.[22] Despite this appeal to American patriotism, the proposal aroused no enthusiasm in Congress, nor did it excite the general public, and in the face of this obvious indifference and the expenditure required to achieve it, the initiative was quietly abandoned.

Given its Cold War origins, the end of the Cold War inevitably raised doubts about the continuing viability of the space station project. However, the new era created the potential for a far better relationship between Russia and America than had prevailed during the Cold War. The two countries quickly looked for practical and symbolic ways in which this better relationship might be demonstrated. As so often during the previous thirty years, space exploration suggested itself as an effective way to achieve both objectives.

In June 1992 Presidents Bush and Yeltsin signed a new bilateral space cooperation agreement.[23] Article I of the agreement identified a wide range of areas of space cooperation, including joint Russian–American flights on the space shuttle and the Russian *Mir* space station. The framework created by this agreement would make possible a much closer working relationship between the two national space agencies, and pave the way for Russia to join the international space station project. Indeed, the new relationship may well have saved both the Russian and American space stations from being abandoned.[24] Russia was in any case now in a position where its space programme was absolutely dependent upon cooperation with other countries, since the Ukraine and Kazakstan, formerly parts of the Soviet Union, were now independent. Russian space station communication was run from Ukraine and its main launch site was in Kazakstan. The potential difficulties this could produce were shown when, following a political disagreement with Russia, Ukraine cut off communications with Russia's Mir space station.[25]

When President Bill Clinton entered office, he reaffirmed American support for the space station, called for a further redesign to reduce costs, and declared that 'full consideration would be given to the use of Russian assets during the station redesign process'.[26] Russian space officials suggested merging the planned Mir-2 Russian plans with the NASA space station proposals as a way of massively reducing programme costs.[27] Clinton met with Yeltsin at the Vancouver Summit in April 1993 and looking for ways to dramatise the new relationship between the countries, saw a partial merger of the two space programmes as 'the ideal initiative'.[28] However, agreement was delayed for some months because of a problem caused by Russia's cooperation with India's space programme. Russia was in the process of selling cryogenic engine technology to India, but the United States opposed this as part of its global efforts to limit the proliferation of ballistic missile technology. The issue was successfully resolved in July 1993, and vice-President Gore and Russian Prime Minister Chernomyrdin signed the agreement bringing Russia in as a space station partner. As a further signal that the Cold War between the two was over, the name of the station was changed from *Freedom* to *Alpha*. Given the long history of rivalry and suspicion between the two countries, the agreement was a historic achievement.[29] It was an important part of the redefinition of the relationship between the two countries that both hoped would shape the new era. It allowed a return to the idealistic rhetoric that had been briefly fashionable at the start of the space age, with NASA Director Goldin declaring that 'as we leave a century full of war, where nations worked together on mega projects that developed weapons, we have an opportunity on the largest project in history, to bring nations together for peaceful cooperation, to signal a new era'.[30]

Idealism certainly lay behind the project, but so did hard political and national security considerations. Russian involvement would allow its struggling space programme to continue and possibly flourish. This was important because the space programme had been one of Russia's few successes in recent history and, for democracy to take root in Russia, the overwhelming sense of national humiliation that had followed the end of the Cold War needed to be countered with some visible successes. In addition, in order to participate, Russia had to join the ballistic missile international control regime led by the United States, and a flourishing space programme made it less likely that Russian rocket scientists would seek employment abroad in countries hostile to the United States. Finally, Russia and the European Space Agency had been discussing the possibility of turning the Mir-2 into a joint Russo–European space station. This would have undermined America's ability to develop the Freedom station on its own, and would have left Russia and Europe operating the world's only space station, a scenario that the United States was anxious to prevent.[31] As early as 1983, NASA administrator Beggs had told his staff that 'if we can attract international cooperation, then other nations will be cooperating with us in the resources that they spend on space, rather than competing with us'.[32] In the event ESA agreed its involvement in the new project at the ESA Council in Toulouse in November 1993. Every ESA member state except the UK signed the agreement. The formal invitation to Russia was made the following month.[33]

The creation of the new international consortium did not bring an end to political difficulties between the partners. The US Iran Non-proliferation Act caused problems, for example. Under the Act the United States is forbidden to make ISS-related payments to Russia unless the President determines that Russia is taking steps to prevent the proliferation of weapons of mass destruction, missile technology and other advanced military systems to Iran.[34] After the loss of the Shuttle *Columbia* in 2003, NASA was prevented from funding Russian Soyuz and Progrez spacecraft missions to service the ISS, because of the terms of the Iran Act. Only a great deal of lobbying by NASA and the White House led the Senate to relax the restrictions, in order to allow NASA to use the Russian spacecraft.[35]

The international space station is as much a purely political creation as the Vostok and Apollo programme's were. Reimar Lust, former Director-General of the European Space Agency, declared in 1995, that the ISS had only been developed at all because of political considerations, that the station 'can only be justified in political terms, not really by itself. No convincing concept for its utilization has as yet been developed'.[36] Given this, it is all the more striking that to date the cooperative international effort represented by the ISS 'has proven to be remarkably resistant to the shifting moods of bilateral and international relations'.[37]

The impact of China and Japan

From the late 1980s, the Chinese economy began to undergo a remarkable transformation so that by the mid-1990s it was producing double-digit economic growth annually, a pattern it has maintained for over a decade, in the process transforming China's capabilities and potential. The government was able to increase dramatically investment in many areas, notably including the space programme, and this allied to the increasing importance of prestige questions led China to re-emphasise development of a manned space programme.

In 2003 China orbited its first astronaut, Yang Liwei in the Shenzou 4 spacecraft, and two years later sent a crew of two into orbit in Shenzhou 6 on a five-day mission. The manned programme is an impressive achievement for China, making her only the third country in the world to launch her own astronauts into space on her own spacecraft. What is equally striking, however, is the dramatic impact that the Chinese achievement has had on other space powers. The 2003 Chinese manned flight was a watershed in terms of re-energising other countries who felt compelled to compete with China, triggering a new, prestige-driven space race.

In April 2005 the Japanese space agency, JAXA, announced new, ambitious plans for a manned space programme, including a landing on the Moon by

2025. This marked a major change in Japanese space policy that had been signalled the previous year, when a Japanese government panel had called for a reorientation of the space programme from unmanned scientific planetary probes to manned space missions. Related plans to establish a permanent base on the Moon would require a six-fold increase in the Japanese space budget, which is only about one-tenth of NASA's.[38] The dramatic announcement followed nearly a decade of low activity and set-backs in the Japanese space programme. While Japan had included plans for a manned spacecraft in earlier space plans, no priority had been given to the programme, which had therefore failed to materialise. The acceleration of the Chinese space programme, to include manned missions to the Moon and eventually beyond, clearly acted as a political spur to the Japanese government. The only other element in Japanese space activity which has accelerated in recent years is the military reconnaissance programme, largely directed at North Korea. Manned missions contribute nothing to military security, and would clearly divert funding from the military programme. That Japan has chosen to pursue a manned programme at this time therefore is a measure of the degree to which it feels that the Chinese manned programme is a direct challenge to Japan's prestige and relative political standing in the region. China is unlikely to be prepared to let its advantage slip. 'Given the strong anti-Japanese sentiment among the Chinese public, and the frosty relations at government level, there is little doubt that China's National Space Administration will not want to be upstaged by its Japanese counterpart'.[39] Effectively the two Asian states have entered a new space race, and they are not the only Asian entrants.

The dramatic announcement by India of a manned programme has been discussed earlier. India has insisted that its manned programme was a logical next step for the Indian programme, and not simply a reaction to the emergence of the Chinese manned programme.[40] That may well be, but the shift in policy is so dramatic in comparison to the almost ideological previous opposition to manned flights, that the timing seems hardly coincidental. As with Japan it may be that the Chinese launches were the ultimate spur necessary to accelerate what otherwise might have been a much longer-term development.

Nor were the Asian states necessarily the only ones for whom China's progress in space raised major questions. On January 2004, shortly after the first Chinese manned flight, President George W. Bush announced a dramatic new programme for American manned exploration of the Solar System. The plans announced included the completion of the International Space Station by 2010, the development of a new manned spacecraft to explore beyond Earth orbit, manned landings on the Moon by 2015 'with the goal of living and working there for extended periods' and preparations for manned missions to Mars.[41]

Fifty years after the launch of Sputnik, the issue of prestige and international perceptions of political leadership have returned to dominate the agenda of space exploration. During the first five decades of the Space Age, politics were always at the heart of space activity, and this remains the case. What has been striking about the international politics of space has been their lack of novelty. To date they have precisely mirrored terrestrial preoccupations and approaches. This is as true of the adversarial elements as it is of the cooperative ones. There are powerful pressures encouraging international cooperation in the utilisation of space, but there are equally powerful nationalist pressures to act alone, and a complex set of factors determines which will prevail in particular relationships and historical periods.

In one sense space can be seen as a 'final frontier' to be crossed, but in another sense humanity has brought its frontiers with it into space, replicating the political divisions and tensions that characterise global politics. The movement into space has also profoundly affected terrestrial politics because of the impact it has had on the security, military, economic, environmental and cultural dimensions of life on Earth. Yet it has also been a vehicle for the creation of complex epistemic communities and international organisations. It still remains a distinctive arena nevertheless, because however much human activities there tend to reflect terrestrial realities, it continues to encourage international actors to believe that it *ought* to be possible to do things differently, and better, beyond the security of our home planet.

Notes

Introduction

1 Walter A. McDougall, *The Heavens and the Earth: A Political History of the Space Age* (New York, Basic Books, 1985), p. 178.
2 Andrew Smith, *Moondust* (London, Bloomsbury, 2005), p. 273.
3 Bram Groen and Charles Hampden-Turner, *The Titans of Saturn: Leadership and Performance Lessons from the Cassini-Huygens Mission* (Singapore, Marshall Cavendish, 2005), p. xvii.
4 Columba Peoples, 'Haunted dreams: critical theory and the militarisation of space', Paper presented at the Annual Conference, British International Studies Association (Cork, December, 2006), p. 22.

1 Perceptions of space and international political theory

1 Bram Groen and Charles Hampden-Turner, *The Titans of Saturn* (Singapore, Marshall-Cavendish, 2005), p. 23.
2 Robert Cox, 'Social forces, states and world orders: beyond international relations theory', in Robert O. Keohane (ed.), *Neorealism and Its Critics* (New York, Columbia University Press, 1986), p. 207.
3 Bram Groen and Charles Hampden-Turner, *The Titans of Saturn* (Singapore, Marshall-Cavendish, 2005), p. 23.
4 Groen and Hampden-Turner, *The Titans of Saturn*, p. 2.
5 Jack Manno, *Arming the Heavens: The Hidden Military Agenda for Space 1945–1995* (New York, Dodd, Mead and Co., 1984), p. 23.
6 Michael Sheehan, *Arms Control: Theory and Practice* (Oxford, Basil Blackwell, 1988), p. 24.
7 Robert S. Cooper, 'No sanctuary: a defense perspective on space', *Issues in Science and Technology*, Vol. 2, No. 3 (Spring, 1986). Reprinted in US Department of Defense, *Earlybird: Special Edition* (Washington, DC, 17 June 1986), p. 18.
8 James Killian, *Sputniks, Scientists and Eisenhower: A Memoir of the First Special Assistant to the President for Science and Technology* (Cambridge, MA: MIT Press, 1997), p. 2.
9 Walter A. McDougall, *The Heavens and the Earth: A Political History of the Space Age* (New York, Basic Books, 1985), p. 418.
10 Andrew Smith, *Moondust* (London, Bloomsbury, 2005), p. 57.
11 McDougall, *Heavens and the Earth*, p. 414.
12 Jurgen Scheffran, 'Peaceful and sustainable use of space – principles and criteria for evaluation', in Wolfgang Bender, Regina Hagen, Martin Kalinowski and

Jurgen Scheffren (eds), *Space Use and Ethics*, Vol. I (Münster, Agenda-Verlag, 2001), p. 50.

13 Richard N. Lebow, 'Classical realism', in Tim Dunne, Milja Kurki and Steve Smith (eds), *International Relations Theories: Discipline and Diversity* (Oxford, Oxford University Press, 2007), p. 57.

14 John Herz, 'Idealist internationalism and the security dilemma', *World Politics*, Vol. 2 (1950), p. 157.

15 John J. Mearsheimer, 'Structural Realism', in Tim Dunne, Milja Kurki and Steve Smith (eds), *International Relations Theories: Discipline and Diversity* (Oxford, Oxford University Press, 2007), p. 74.

16 Joseph M. Goldsen (ed.), *Outer Space in World Politics* (London, Pall Mall Press, 1963), p. 4.

17 Klaus Knorr, 'The international implications of outer space activities', in Joseph M. Goldsen (ed.), *Outer Space in World Politics* (London, Pall Mall Press, 1963), p. 117.

18 Joan Johnson-Freese and Andrew S. Erickson, 'The emerging China–EU space partnership: a geotechnological balancer', *Space Policy*, Vol. 22 (2006), p. 13.

19 Johnson-Freese and Erickson, 'The emerging China–EU space partnership', p. 13.

20 Eric R. Sterner, 'International competition and co-operation: civil space programmes in transition', *The Washington Quarterly*, Vol. 16 (1993), p. 129.

21 John Spanier, *Games Nations Play* (New York, Holt, Reinhart & Winston, 1984), p. 120.

22 Don E. Cash, *The Politics of Space Cooperation* (West Lafayette, IN, Purdue University Press, 1967), p. 130.

23 Raymond Aron, *Peace and War: A Theory of International Relations* (London, Weidenfeld and Nicolson, 1966), p. 658.

24 Robert J. Mrazek, 'Rethinking national and global security – the role of space-based observations', *Space Policy*, Vol. 5 (1989), p. 156.

25 Mearsheimer, 'Structural realism', p. 72.

26 Steve Weber, 'Realism, détente and nuclear weapons', *International Organisation*, Vol. 44 (1990), pp. 58–9.

27 Robert Dallek, *John F. Kennedy: An Unfinished Life* (London, Penguin, 2004), p. 654.

28 Charles Glaser, 'Realists as Optimists: cooperation as self-help', *International Security*, Vol. 19 (1994–5), pp. 50–90.

29 Jack Donnelly, 'Realism', in Scott Burchill, Andrew Linklater, Richard Devetak, Jack Donnelly, Matthew Paterson, Christian Reus-Smit and Jacqui True, *Theories of International Relations* (3rd edn) (London, Palgrave-Macmillan, 2005), p. 31.

30 John Spanier, *Games Nations Play*, p. 120.

31 Robert J. Mrazek, 'Rethinking national and global security – the role of space-based observations', *Space Policy*, Vol. 5 (1989), pp. 158–9. This possibility had been suggested even earlier by Curien, see, Hubert Curien, 'For peace or for war? Competition in the control of outer space', *NATO's Fifteen Nations*, Vol. 27 (April–May, 1982), p. 20.

32 Herman Pollack, 'International relations in space – a US view', *Space Policy*, Vol. 4 (1988), pp. 24–5.

33 Thomas E. Cremins, 'Security in the space age', *Space Policy*, Vol. 6 (1990), pp. 33–5.

34 Hubert Curien, 'For peace or for war? , p. 18.

35 Lisa A. Martin, 'Neoliberalism', in Tim Dunne, Milja Kurki and Steve Smith (eds), *International Relations Theories: Discipline and Diversity* (Oxford, Oxford University Press, 2007), p. 111.

36 Lisa Martin, 'Neoliberalism', p. 111.

37 Allan M. Din, 'Stopping the arms race in outer space', *Journal of Peace Research*, Vol. 20 (1983), pp. 223–5.

38 For earlier calls for such a regime see, Klaus Knorr, 'The international implications of outer space activities', in Joseph M.Goldsen (ed.), *Outer Space in World Politics* (London, Pall Mall Press, 1963), p. 119; Michael Sheehan, *The Arms Race* (Oxford, Martin Robertson, 1983), p. 107.

39 Klaus Knorr, 'The international implications of outer space activities', pp. 118–19.

40 Burchill, p. 72.

41 Richard Garwin, 'Space defense – the impossible dream?', *NATO's Sixteen Nations* (April, 1986), p. 23.

42 Nicholas Onuf, *World of Our Making: Rules and Rule in Social Theory and International Relations* (Columbia, SC, University of South Carolina Press, 1989).

43 K. M. Fierke, 'Constructivism', in Tim Dunne, Milja Kurki and Steve Smith (eds), *International Relations Theories: Discipline and Diversity* (Oxford, Oxford University Press, 2007), p. 168.

44 Jurgen Scheffren, 'Peaceful and sustainable use of space – principles and criteria for evaluation', in Wolfgang Bender, Regina Hagen, Martin Kalinowski and Jurgen Scheffren (eds), *Space Use and Ethics*, Vol. I (Münster, Agenda-Verlag, 2001), p. 51.

45 David Harvey, *The Condition of Postmodernity* (Oxford, Blackwell, 1990), p. 249.

46 Bettyann Holtzmann Kelvas, *Almost Heaven: The Story of Women in Space* (Cambridge, MA, MIT Press, 2005), p. x.

47 A phrase used by David J. Shayler and Ian Moule, *Women in Space: Following Valentina* (Chichester, Springer Praxis, 2005), p. 70.

48 Paul G. Dembling, 'Commercial utilisation of space and the law', *Yearbook of Air and Space Law* (1967), pp. 283–4.

49 J. W. Durch and D. A. Wilkening, 'Steps into space', in J. W. Durch (ed.), *National Interests and the Military Use of Space* (Cambridge, MA, Ballinger Publishing Co., 1984), p. 11.

50 Speech by Senator Lyndon Johnson, Washington DC, 14 January 1958. Cited in OTA, *International Cooperation and Competition in Civilian Space Activities* (Washington, DC, US Congress, Office of Technology Assessment, OTA-ISC-239, July 1985), p. 35.

51 Allan Rosas, 'The militarisation of space and international law', *Journal of Peace Research*, Vol. 20 (1983), p. 357.

52 Jeff Kingwell, 'The militarization of space – a policy out of step with world events', *Space Policy*, Vol. 6 (1990), p. 110.

53 Herman Pollack, 'International relations in space – a US view', *Space Policy*, Vol. 4 (1988), p.26.

54 Herman Pollack, 'International relations in space', p. 26.

55 These issues are discussed in detail in Chapter 7.

2 Propaganda and national interest

1 R. T. Newman, 'Propaganda: an instrument of foreign policy', *Columbia Journal of International Affairs*, Vol. 5 (1951), p. 56.
2 Quoted in A. Frye, 'Soviet space activities: a decade of Pyrrhic politics', in L. P. Bloomfield (ed.), *Outer Space: Prospects for Man and Society* (New York, Frederick Praeger, 1968), p. 193.
3 M. S. Smith, 'Evolution of the Soviet space programme from Sputnik to Salyut and beyond', in U. Ra'anan and R. L. Pfaltzgraff (eds), *International Security Dimensions of Space* (Hamden, CT, Archon Books, 1984), p. 286.
4 K. J. Holsti, *International Politics: A Framework for Analysis* (5th edn) (Englewood Cliffs, NJ, Prentice-Hall, 1988), p. 207.
5 R. E. Lapp, *Man and Space* (London, Secker and Warburg, 1961), p. 15.
6 J. M. Mackintosh, *Strategy and Tactics of Soviet Foreign Policy* (Oxford, Oxford University Press, 1962), p. 274.
7 Quoted in G. A. Skuridin, *Entrance of Mankind into Space* (Washington, DC, National Aeronautics and Space Administration, 1976), p. 4.
8 W. A. McDougall, 'Sputnik, the space race and the Cold War', *Bulletin of the Atomic Scientists*, Vol. 41, No. 5 (1985), p. 20.
9 Michael Sheehan, *Balance of Power: History and Theory* (London, Routledge, 1996), pp. 14–15.
10 Julian Lider, *Correlation of Forces* (Aldershot, Gower, 1986), p. 124.
11 D. Mikheyev, *The Soviet Perspective on the Strategic Defence Initiative* (London, Pergamon-Brasseys, 1987), p. 38.
12 R. D. Humble, *The Soviet Space Programme* (London, Routledge, 1988), p. 1.
13 Peter A. Gorin, 'Rising from the cradle: Soviet Perceptions of space before Sputnik', in Roger D. Launius, John M. Logsdon and Robert W. Smith (eds), *Reconsidering Sputnik: Forty Years Since the Soviet Satellite* (London, Routledge, 2000), p. 18.
14 C. Lee, *War in Space* (London, Hamish Hamilton, 1986), pp. 37–9.
15 J. Manno, *Arming the Heavens: The Hidden Military Agenda for Space, 1945–1995* (New York, Dodd, Mead and Co., 1984), p. 10.
16 C. Lee, *War in Space*, p. 161.
17 J. Popescu, *Russian Space Exploration: The First 21 Years* (Oxford, Gothard House, 1979), p. 1.
18 Peter A. Gorin, 'Rising from the cradle', p. 36.
19 Asif A. Siddiqi, 'Korolev, Sputnik and the IGY', in Roger D. Launius *et al.* (eds), *Reconsidering Sputnik*, p. 49.
20 R. D. Humble, *The Soviet Space Programme* (London, Routledge, 1988), p. 4.
21 J. Oberg, *Red Star in Orbit* (New York, Random House, 1981), p. 29.
22 J. Trux, *The Space Race* (London, New English Library, 1987), p. 13.
23 W. A. McDougall, 'Sputnik, the space race and the Cold War', *Bulletin of the Atomic Scientists*, Vol. 41, No. 5 (1985), p. 22.
24 S. J. Lewis and R. A. Lewis, *Space Resources: Breaking the Bonds of Earth* (New York, Columbia University Press, 1987), p. 20; N. L. Johnson, *Soviet Military Strategy in Space* (London, Jane's, 1987), p. 17.
25 Siddiqi, 'Korolev, Sputnik and the IGY', p. 58.
26 Trux, p. 14.
27 H. Young, B. Silcock and P. Dunn, *Journey to Tranquillity* (London, Jonathan Cape, 1969), p. 68.
28 Peter A. Gorin, 'Rising from the cradle', p. 39.
29 A. Frye, 'Soviet space activities: a decade of Pyrrhic politics', in L. P. Bloomfield (ed.), *Outer Space: Prospects for Man and Society* (London, Praeger, 1968), p.

194; J. M. Goldsen, 'Outer space in world politics', in J. M. Goldsen (ed.), *Outer Space in World Politics* (London, Pall Mall Press, 1963), p. 8.

30 Oberg, *Red Star in Orbit*, p. 33; Asif Siddiqi, *Sputnik and the Soviet Space Challenge* (Gainesville, FL, University Press of Florida, 2003), p. 171.

31 B. Harvey, *Race into Space: The Soviet Space Programme* (Chichester, Ellis Horwood, 1988), p. 32.

32 *Manchester Guardian*, 7 October 1957. Quoted in James J Harford, 'Korolev's triple play', in Roger D. Launius, John M. Logsdon and Robert W. Smith (eds), *Reconsidering Sputnik: Forty Years Since the Soviet Satellite* (London, Routledge, 2000), p. 85.

33 D. Mikheyev, *The Soviet Perspective on the Strategic Defence Initiative*, p. 39.

34 *Pravda*, 26 August 1959.

35 McDougall, *Bulletin*, p. 24.

36 Nicholas L. Johnson, *Soviet Military Strategy in Space* (London, Jane's), p. 18.

37 S. Leskov, 'Soviet space in transit', *Space Policy* (August, 1989), p. 184.

38 Asif Siddiqi, *Sputnik and the Soviet Space Challenge*, p. 171.

39 T. Osman, *Space History* (London, Michael Joseph, 1983), pp. 44–5.

40 McDougall, *Bulletin*, p. 24.

41 D. Mikheyev, p. 38.

42 Russian space historian Georgiy Vetrov has suggested that it was triggered by reports of Senator Lyndon Johnson's remarks to the Senate Democratic Caucus suggesting that 'control of space means control of the world'. See Siddiqi, *Sputnik and the Soviet Space Challenge*, p. 237.

43 F.C.Barghoorn, *Soviet Foreign Propaganda* (Princeton, NJ, Princeton University Press, 1964), p. 200.

44 G. A. Skuridin, *Entrance of Mankind into Space*, p. 2.

45 Barghoorn, p. 201.

46 W. A. McDougall, *The Heavens and the Earth: A Political History of the Space Age* (New York, Basic Books, 1985), p. 288.

47 G. A. Skuridin, *Entrance of Mankind into Space* (Washington, DC, NASA, 1976), p. 31.

48 J. Oberg, *Red Star in Orbit*, pp. 69–70.

49 Siddiqi, *Sputnik and the Soviet Space Challenge*, p. 292.

50 G. Perry, 'Perestroika and Glasnost in the Soviet space programme', *Space Policy* (November, 1989), p. 283.

51 Siddiqi, *Sputnik and the Soviet Space Challenge*, p. 379.

52 J. Logsdon and A. Dupas, 'Was the race to the Moon real?', *Scientific American* (June, 1994), p. 17.

53 N. J. Johnson, *Soviet Military Space Strategy*, p. 9.

54 D. Mkheyev, p. 36.

55 Siddiqi, *Sputnik and the Soviet Space Challenge*, p. 393.

56 Oberg, *Red Star in Orbit*, p. 77.

57 While Khruschev traditionally receives all the blame for this mission, post-Soviet research suggests that in fact Korolev was not opposed to it. Knowing that the Soyuz spacecraft would not be ready before 1965 at the earliest, he was keen to maintain the image of the Soviet space programme consistently achieving objectives before its American rivals and therefore backed Voskod so that the Soviet programme would be seen as active and successful in 1964–5, rather than in a hiatus, waiting for Soyuz to become available. See Siddiqi, *Sputnik and the Soviet Space Challenge*, pp. 383–5.

58 Dodd L. Harvey and Linda C. Ciccoritti, *US–Soviet Cooperation in Space* (Miami, FL, Center for Advanced International Studies, University of Florida, 1974), p. 123.
59 Siddiqi, *Sputnik and the Soviet Space Challenge*, p. 351.

3 The new frontier

1   Alan J. Levine, *The Missile and Space Race* (Westport, CT and London, Praeger, 1994), p. 2.
2   Tony Osman, *Space History* (London, Michael Joseph, 1983), p. 23.
3   Levine, *Missile and Space Race*, p. 3.
4   Levine, *Missile and Space Race*, p. 4.
5   Matt Bille and Erika Lishock, *The First Space Race* (College Station, TX, A&M University Press, 2004), p. 15.
6   Bille and Lishock, *First Space Race*, p. 22.
7   Walter A. McDougall, *The Heavens and the Earth: A Political History of the Space Age* (New York, Basic Books, 1985), p. 101.
8   Bille and Lishock, *First Space Race*, p. 28.
9   RAND Corporation, *Preliminary Design for an Experimental World-Circling Spaceship* (Santa Monica, CA, Project RAND, 1946; reprinted RAND Corporation, 1999), p. 10.
10 RAND, *Preliminary Design*, pp. 9–16.
11 Asif A. Siddiqi, 'Korolev, Sputnik and the International Geophysical Year', in Roger D. Launius, John M. Logsdon and Robert W. Smith (eds), *Reconsidering Sputnik: Forty Years since the Soviet Satellite* (London, Routledge, 2000), p. 48.
12 James J. Harford, 'Korolev's triple play: Sputniks 1, 2 and 3', in Roger Launius *et al.* (eds), *Reconsidering Sputnik*, p. 77.
13 Dwayne A. Day, 'Cover stories and hidden agendas: early American space and national security policy', in Roger Launius *et al.* (eds), *Reconsidering Sputnik*, p. 167.
14 NSC 5520, Comments on the Report to the President by the Technological Capabilities Panel. Quoted in Dwayne A. Day, 'Cover stories', p. 170.
15 Walter A. McDougall, *The Heavens and the Earth*, p. 29.
16 H. Brooks, 'The motivations for space activity', in A. A. Needell (ed.), *The First Twenty-five Years in Space: A Symposium* (Washington, DC, Smithsonian Institution Press, 1983), p. 6.
17 Walter A. McDougall, *The Heavens and the Earth*, pp. 144–5.
18 *New York Times*, 6 October 1957.
19 S. Ramo, 'The practical dimensions of space', in Needell, *The First Twenty-five Years*, p. 51.
20 Gene Kranz, *Failure is Not an Option: Mission Control from Mercury to Apollo 13 and Beyond* (New York, Berkley Books, 2001), p. 15.
21 Saki Dockrill, *Eisenhower's New Look National Security Policy, 1953–61* (London, Macmillan Press, 1996), p. 216.
22 T. A. Heppenheimer, *Countdown: A History of Spaceflight* (New York, John Wiley & Sons, 1997), p. 125.
23 National Security Council, *Preliminary US Policy on Outer Space*, NSC5814/1, 18 August 1958.
24 J. R. Killian, *Sputnik, Scientists and Eisenhower* (Cambridge, MA, MIT Press, 1977), p. xv.

25 P. Stares, *The Militarisation of Space: US Policy 1945 to 1984* (Ithaca, NY, Cornell University Press, 1985), p. 42.

26 W. D. Kay, *Defining NASA: The Historical Debate Over the Agency's Mission'* (Albany, NY, State University of New York Press, 2005), p. 57.

27 Mansfield Sprague, 'Proposal for a national security policy on outer space', International Security Affairs, Assistant Secretary of Defense, 25 February 1958. Cited in McDougall, *The Heavens and the Earth*, p. 178.

28 J. Logsdon, 'Introduction', in Needell, *The First Twenty-five Years*, p. 4.

29 E. M. Emme (ed.), *The History of Rocket Technology* (Detroit, MI, Wayne State University Press, 1964), p. 434.

30 *National Aeronautics and Space Act*, 85th Congress, 29 July 1958.

31 Bram Groen and Charles Hampden-Turner, *The Titans of Saturn: Leadership and Performance Lessons from the Cassini–Huygens Mission* (Singapore, Marshall Cavendish, 2005), p. 25.

32 Dwayne A. Day and Colin Burgess, 'Monkey in a blue suit', *Spaceflight*, Vol. 48, No. 7 (July, 2006), p. 266.

33 George M. Moore, Vic Budura and Joan Johnson-Freese, 'Joint space doctrine: catapulting into the future', *Joint Forces Quarterly* (Summer, 1994), p. 74.

34 K. Hufbauer, 'Solar observational capabilities and the solar physics community since Sputnik', in M. Collins and S. Fries, *A Spacefaring Nation* (Washington, DC, Smithsonian Institution Press, 1991), p. 78.

35 S. Ramo, *The Practical Dimensions*, p. 60.

36 *Newsweek* (14 October, 1957), p. 37.

37 Jeremy Isaacs and Taylor Downing, *Cold War* (London, Bantam Press, 1998), p. 156.

38 Roger Hanberg, 'Rationales of the space program', in Eliger Sadeh (ed.), *Space Politics and Policy: An Evolutionary Perspective* (Dordrecht, Kluwer Academic Publishers, 2002), p. 34.

39 A. Etzioni, 'Comments', in Needell (ed.), *The First Twenty-five Years*, pp. 33–6.

40 Walter A. McDougall, 'Scramble for space', *Wilson Quarterly*, Vol. 4 (Autumn, 1980), p. 59.

41 David Farber (ed.), *The Sixties: From Memory to History* (London, University of North Carolina Press, 1994), p. 19.

42 John Noble Wilford, 'Riding high', *The Wilson Quarterly* (Autumn, 1980), p. 64.

43 President's Science Advisory Committee, *Report of the Ad-hoc Committee on Man-in-Space*, 16 December 1960. Quoted in W. D. Kay, *Defining NASA*, p. 60.

44 Robert Parkinson, *Citizens of the Sky* (Stotfold, 2100 Publishing, 1987), p. 13.

45 Charles Murray and Catherine Cox, *Apollo: The Race to the Moon* (London, Secker & Warberg, 1989), p. 67.

46 President John F. Kennedy, text of address to Congress entitled, 'Special message to the Congress on urgent national needs', 25 May 1961. John F. Kennedy Library, http://www.cs.umb.edu/jfklibrary/j052561.htm.

47 President Kennedy, special message to Congress, 25 May 1961.

48 Theodore Sorenson, *Kennedy* (London, Hodder & Stoughton, 1965), pp. 524–6.

49 William L. O'Neill, *Coming Apart: An Informal History of America in the 1960's* (New York, Time Books/Random House, 1971), p. 59.

50 James Schefter, *The Race: The Complete Story of How America Beat Russia to the Moon* (New York, Anchor Books, 2000), p. 143.

51 Edmund S. Ions, *The Politics of John F. Kennedy* (London, Routledge and Kegan Paul, 1967), p. 51.
52 President John F. Kennedy, Address before the 18th General Assembly of the United Nations, New York, 20 September, 1963. John F. Kennedy Library, www.cs.umb.edu/jfklibrary/j092063.htm.
53 Tim Furniss, *One Small Step: The Apollo Missions, The Astronauts, the Aftermath – A Twenty Year Perspective* (Yeovil, Haynes Publishing Group, 1989), pp. 11–12.
54 Murray and Cox, *Apollo: The Race to the Moon*, p. 77.
55 Theodore R. Simpson (ed.), *The Space Station: An Idea Whose Time Has Come* (New York, IEEE Press, 1985), pp. 54–5.
56 Murray and Cox, p. 77.
57 W. D. Kay, *Defining NASA*, pp. 68–9.
58 Buzz Aldrin and Malcolm McConnell, *Men from Earth* (London and New York, Bantam Press, 1989), p. 62.
59 Robert A. Divine, *The Johnson Years, Volume 2* (Lawrence, KS, University Press of Kansas, 1987), p. 223.
60 Michael R. Beschlass, 'Kennedy and the decision to go to the moon', in Roger D. Launius and Howard E. McCurdy (eds), *Spaceflight and the Myth of Presidential Leadership* (Champaign, IL, University of Illinois Press, 1997), p. 57.
61 Murray and Cox, p. 79.
62 W. D. Kay, *Defining NASA*, pp. 73–4.
63 Trumbull Higgins, *The Perfect Failure: Kennedy, Eisenhower, and the CIA at the Bay of Pigs* (New York, W. W. Norton, 1987), p. 67.
64 Higgins, *Perfect Failure*, p. 150.
65 Murray and Cox, p. 80.
66 J. Logsdon, 'Evaluating Apollo', *Space Policy*, Vol. 5, No. 3 (August, 1989), p. 188.
67 Howard McCurdy, 'The decision to build the space station: too weak a commitment?', *Space Policy*, Vol. 4, No. 4 (November, 1988), p. 298.
68 Murray and Cox, p. 83.
69 Letter from NASA Administrator Webb and Secretary of Defense McNamara to Vice President Johnson, 8 May 1961. Quoted in W. D. Kay, *Defining NASA*, p. 75.
70 Murray and Cox, p. 74.
71 David Callahan and Fred I. Greenstein, 'The reluctant racer: Eisenhower and US space policy', in Roger D. Launius and Howard E. McCurdy (eds), *Spaceflight and the Myth of Presidential Leadership* (Champaign, IL, University of Illinois Press, 1997), p. 43.
72 James Schefter, *The Race*, p. 143.
73 Trux, p. 30.
74 Logsdon, *Evaluating Apollo*, p. 190.
75 Peter Fairley, *Man on the Moon* (London, Mayflower Books, 1969), p. 72.
76 Howard McCurdy, *The Space Station Decision: Incremental Politics and Technological Choice* (Baltimore, MD, Johns Hopkins University Press, 1990), p. 21.
77 Fairley, *Man on the Moon*, p. 75.
78 McCurdy, *The Space Station Decision*, p.17.
79 'Towards a new era in space: realigning US policies to new realities', Committee on Space Policy, National Academy of Sciences/National Academy of Engineering, *Space Policy*, Vol. 5, No. 3 (August 1989), p. 237.
80 Andrew Smith, *Moondust* (London, Bloomsbury, 2005), p. 204.

81 Everett C. Dolman, *Astropolitik: Classical Geopolitics in the Space Age* (London, Frank Cass, 2002), p. 15.
82 President John F. Kennedy, Address to the UN General Assembly, 20 September 1963.
83 Walter A. McDougall, *The Heavens and the Earth*, p. 108.
84 William L. O'Neill, *Coming Apart*, p. 57.
85 Roger D. Launius, 'Historical dimensions of the space age', in Eligar Sadeh (ed.), *Space Politics and Policy: An Evolutionary Perspective* (Dordrecht, Kluwer Academic Publishers, 2002), p. 3.
86 William L. O'Neill, *Coming Apart*, p. 58.
87 Chris Kraft, *Flight: My Life in Mission Control* (New York, Plume/Penguin Books, 2002), p. 317.
88 Herman Pollack, 'Impact of the space program on America's foreign relations', in US Congress, Senate, Committee on Aeronautical and Space Sciences, *International Cooperation in Outer Space: A Symposium*, Senate Document no 92-57, 92nd Congress, 1st session (Washington, DC, US Government Printing office, 1971), p. 601.

4 International cooperation in space

1 OTA, *International Cooperation and Competition in Civilian Space Activities* (Washington, DC, US Congress, Office of Technology Assessment, OTA-ISC-239, July, 1985), p. 41.
2 V. Sevastyanov and A. Ursol, 'Cosmonautics and social development', *International Affairs* (Moscow), November, 1977, p. 72.
3 *US Foreign Broadcast Information Service, Daily Report. Soviet Union*, Vol. 3, No. 36, Supplement 1 (24 February 1981), p. 6.
4 OTA, *International Cooperation*, p. 43.
5 Peter Bond, *The Continuing Story of the International Space Station* (Chichester, Springer Praxis, 2002), p. 44.
6 Peter Bond, *The Continuing Story*, p. 45.
7 Sevastyanov and Ursol, p. 75.
8 Seyastyanov and Ursol, p. 71.
9 Seyastyanov and Ursol, p. 71.
10 *Soviet Space Programs 1976–80, Part I*, Committee on Commerce, Science and Transportation, US Senate, 97th Congress, 2nd Session (Washington, DC, US Government Printing Office, 1982), p. 177
11 J. Downing, 'Cooperation and competition in satellite communication: the Soviet Union', in D. Demac (ed.), *Tracing New Orbits – Cooperation and Competition in Global Satellite Development* (New York, Columbia University Press, 1986), pp. 284–5.
12 *Soviet Space Programs 1976–80, Part I*, Committee on Commerce, Science and Transportation, US Senate, 97th Congress, 2nd Session (Washington, DC, US Government Printing Office, 1982), p. 288.
13 Brian Harvey, *The Japanese and Indian Space Programmes: Two Roads into Space* (Chichester, Springer Praxis Books, 2000), p. 164.
14 Krasnaya Zvesda (Red Star), 12 April 1977. Quoted in Y. Karash, *The Superpower Odyssey: A Russian Perspective on Space Cooperation* (Reston, VA, American Institute of Aeronautics and Astronautics, 1999), p. 130.
15 J. Oberg, *The New Race for Space* (Harrisburg, PA, Stackpole Books, 1984), p. 3.
16 Oberg, *The New Race for Space*, p. 3.

17 Karash, *The Superpower Odyssey*, p. 75.
18 Alexandr Kazantsev, *Lunar Road*; and Georgiy Martynov, *220 Days in a Starship*. Cited in Peter Gorin, 'Rising from the cradle: Soviet perceptions of space flight before Sputnik', in Roger D. Launius, John M. Logsdon and Robert W. Smith (eds), *Reconsidering Sputnik: Forty Years Since the Soviet Satellite* (London, Routledge, 2000), p. 32.
19 W. D. Kay, *Defining NASA: The Historical Debate Over the Agency's Mission* (Albany, NY, State University of New York Press, 2005), p. 84.
20 *Congressional Record*, 10 October 1963, cited in Kay, *Defining NASA*, p. 85.
21 President Lyndon Johnson, Press Conference, 29 August 1965. Quoted in Kay, *Defining NASA*, p. 96.
22 OTA, *International Cooperation*, p. 410.
23 Yuri Karash, p. 73.
24 Matthew J. von Bencke, *The Politics of Space: A History of US–Soviet Competition and Cooperation* (Boulder, CO, Westview, 1997); von Benke, *The Politics of Space*, pp. 79–80.
25 von Bencke, *The Politics of Space*, p. 79.
26 *Pravda*, 6 June 1971, quoted in Karash, p. 105.
27 *Congressional Record*, 5 June 1972, S.8782. Quoted in Karash, p. 113.
28 *Moscow TASS*, International Service in English, 26 May 1972. Quoted in Karash, p. 113.
29 OTA, *International Cooperation*, p. 377.
30 Peter Bond, *The Continuing Story*, p. 103.
31 Hans Mark, *The Space Station: A Personal Journey* (Durham, NC, Duke University Press, 1987), p. 50.
32 James Oberg, 'Russia's space program running on empty', http://astronautix.com/articles/ruspart2.htm.
33 Committee on Commerce, Science and Transportation, United States Senate, *Soviet Space Programs 1976–80, Part I* (Washington, DC, US Government Printing Office, 1982), p. 14.
34 M. Walker, *The Waking Giant: The Soviet Union Under Gorbachev* (London, Michael Joseph, 1986), p. 118.
35 D. Mikheyev, *The Soviet Perspective on the Strategic Defence Initiative* (London, Pergamon-Brasseys, 1987), p. 3.
36 Vernon van Dyke, *Pride and Power: The Rationale of the Space Program* (London, Pall Mall Press, 1965), p. 232
37 OTA, *International Cooperation*, p. 33.
38 Quoted in Vernon van Dyke, *Pride and Power*, p. 235.
39 T. Keith Glennan, *Statement before the Institute of World Affairs* (Pasadena, CA, 7 December 1961). Quoted in *United States Civilian Space Programmes 1958–1978*, Report prepared for the Subcommittee on Space Science and Applications, Committee on Science and Technology, House of Representatives, 97th Congress, first session, Volume I (Washington, DC, US Government Printing Office, 1981), p. 835.
40 Don E. Cash, *The Politics of Space Cooperation* (West Lafayette, IN, Purdue University Press, 1967), p. 128.
41 *Senate Resolution 327, Report No. 1925*, 85th Congress, 2nd session, 24 July 1958.
42 *National Aeronautics and Space Act*, 1958, section 205.
43 International Aspects of the National Aeronautics and Space Act of 1958, as amended (72 stat. 426). Quoted in OTA, *International Cooperation*, p. 36.
44 Quoted in Vernon van Dyke, *Pride and Power*, p. 245.

45 OTA, *International Cooperation*, p. 41.
46 OTA, *International Cooperation*, p. 257.
47 *Treaty on Principles Governing the Activities of States in the Exploration and Use of Outer Space, Including the Moon and other Celestial Bodies* (New York, United Nations Organization, 1967).
48 M. N. Golovine, *Conflict in Space: A Pattern of War in a New Dimension* (London, Temple Press Limited, 1962), p. 106.
49 Quoted in Vernon van Dyke, *Pride and Power*, p. 60.
50 US Congress, Senate, Committee on Aeronautical and Space Sciences, *Documents on International Aspects of the Exploration and Use of Outer Space, 1954–1962*, Staff Report, S.Doc, No. 18, 88th Congress, 1st session, 1963, p. 226. Quoted in Vernon van Dyke, *Pride and Power*, p. 249.
51 Peter Bond, *The Continuing Story*, p. 71.
52 Eligar Sadeh, James P. Lester and Willy Z. Sadeh, 'Modelling international cooperation for space exploration', *Space Policy*, Vol. 12 (1996), pp. 207–23.
53 Green and Hampden-Turner, *The Titans of Saturn*, p. 1.

5 European integration and space

1 J. Dougherty and R. Pfaltzgraff, *Contending Theories of International Relations* (New York, Harper and Rowe, 1981), p. 419.
2 J. W. Spanier, *Games Nations Play* (New York, Praeger, 1978), p. 533.
3 David Mitrany, 'The functional approach in historical perspective', *International Affairs* (July, 1971), p. 538.
4 Karen Mingst, 'Functionalist and regime perspectives: the case of Rhine river cooperation', *Journal of Common Market Studies*, Vol. 20 (1981), p. 161.
5 R. Lust, 'European cooperation in space', *ESA Bulletin*, No. 58 (Noordwijk, Netherlands, ESA, 1989), p. 19.
6 Spanier, p. 537.
7 Kazuto Suzuki, *Policy Logics and Institutions of European Space Collaboration* (Aldershot, Ashgate, 2003), pp. 2–5.
8 Suzuki, p. 5.
9 J. Krige, 'Europe into space: the Auger years 1959–1967', *ESA HSR-8* (Noordwijk, Netherlands, ESA, 1983), p. 4.
10 *Foreign Relations of the United States 1955–57, Vol.IV, Western European Security and Integration* (Washington, DC, US Government Printing Office, 1986), pp. 218–60.
11 P. Fischer, 'The origins of the Federal Republic of Germany's space policy 1959–1965 – European and national dimensions', *ESA HSR-12* (Noordwijk, Netherlands, ESA, 1994), p. 6.
12 J. Krige, 'The prehistory of ESRO 1959/60', *ESA HSR-1* (Noordwijk, Netherlands, ESA, 1992), p. 5.
13 J. Krige, 'Europe into Space', *ESA HSR-8*, p. 5.
14 Royal Institute for International Affairs, *Europe's Future in Space* (London, Routledge and Kegan Paul, 1988), p.72.
15 J. Krige, *The Auger Years*, p. 11.
16 J. Krige, *The Auger Years*, p. 6.
17 R. Collino, 'The US space program: an international viewpoint', *International Security*, Vol. 11 (1987), p. 159.
18 Fischer, p. 28.
19 *Six* referring to the EEC states and *Seven* to the European Free Trade Association (EFTA) countries. Krige, ELDO, p. 19.

20 Recommendation 251 of 24 September 1960. Quoted in A. H. Robertson, *European Institutions* (London, Stevens & Sons, 1973), p. 259.
21 Robertson, p. 259.
22 J. Krige, *The Auger Years*, p. 6.
23 J. Krige, *The Auger Years*, p. 11.
24 Suzuki, 2003, p. 52.
25 J. Krige, *Launch of ELDO*, p. 18.
26 Fischer, p. 16.
27 Fischer, p. 22.
28 J. Krige, *The Auger Years*, p. 14.
29 Telegram 3497 from British Foreign Office to Washington, 12 August 1960. Cited in J. Krige, 'The launch of ELDO', *ESA HSR-7* (Noordwijk, Netherlands, ESA, 1993), p. 11.
30 Fischer, p. 45.
31 J. Hoagland, 'The other space powers: Europe and Japan', in U. Ra'anan and R. Pfaltzgraff (eds), *International Security Dimensions of Space* (Hamden, CT, Archon Books, 1984), p. 175.
32 P. Geens, 'The new European Space Agency', *ESA Bulletin*, No. 4 (February, 1976), p. 5.
33 K. Madders and W. Thiebault, 'Two Europe's in one space: the evolution of relations between the European Space Agency and the European Community in space affairs', *Journal of Space Law*, Vol. 20, No. 2 (1992), p. 119.
34 Kevin Madders, *A New Force at a New Frontier* (Cambridge, Cambridge University Press, 1997), pp. 131–2.
35 J. Grey, 'The international team', in J. Grey (ed.), *Beachheads in Space* (New York, Macmillan, 1983), p. 49.
36 Charles Sheldon (Chief of Science Policy, Research Division, US Library of Congress), in Grey, p. 48.
37 RIIA, *Europe's Future in Space*, p. 72.
38 *European Yearbook*, Vol.XVI (1968), pp. 815–19. Cited in Robertson, p. 261.
39 *ESA Bulletin*, No. 58 (May 1989), p. i.
40 Stacia E. Zabusky, *Launching Europe: An Ethnography of European Cooperation in Space Science* (Princeton, NJ, Princeton University Press, 1995), p. 49.
41 Zabusky, p. 55.
42 E. Quistgaard, 'ESA and Europe's future in space', *ESA Bulletin*, No. 39 (August 1984), p. 22.
43 Zabusky, 1995, p. 5.
44 Norman Longden and Duc Guyenne (eds), *Twenty Years of European Cooperation in Space: An ESA Report* (Noordwijk, Netherlands, ESA Scientific and Technical Publications Branch, 1984), p. 229.
45 Longden and Guyenne, 1984, p. 229.
46 Suzuki, 2003, p. 103.
47 R. Chipman, 'Multilateral intergovernmental co-operation', in *The World in Space* (Englewood Cliffs, NJ, Prentice Hall, 1982), p. 587.
48 G. Geens, 'The new European Space Agency', *ESA Bulletin*, No. 4 (February 1976), p. 4.
49 R. Lust, 'The long-term European space plan: a basis for cooperation', *ESA Bulletin* (Noordwijk, Netherlands, 1989), p. 11.
50 Kevin Madders, *A New Force*, p. 173.
51 Raanan and Pfaltzgraff, p. 177.
52 Raanan and Pfaltzgraff, p. 178.
53 *ESA Convention*, Article VII.

54　M. Schwartz, 'The politics of European space collaboration', *Intermedia*, Vol. 9, No. 4 (July 1981), p. 69.
55　AWST, 'Europeans looking beyond Ariane 4 to goal of independence in space', *Aviation Week and Space Technology* (9 June 1986), p. 87.
56　Madders and Thiebault, p. 120.
57　Suzuki, 2003, p. 97.
58　Treaty Establishing the European Coal and Steel Community (Paris, 18 April 1951), available at www.ena.lu/mce.cfm.
59　WEU Assembly, *Proceedings*, 13th Ordinary session (1966), Vol. 2 (Documents), Doc 389, pp. 152–3.
60　Madders, *A New Force*, p. 569.
61　C. Layton (EC Director responsible for aerospace industry), 'The European space effort in the light of global European policy', *ESA Bulletin*, No. 4 (1976), p. 32.
62　Layton, p. 33.
63　RIIA, *Europe's Future in Space*, p. 83.
64　Madders and Thiebault, pp. 127–8.
65　Madders, *A New Force*, p. 570.
66　European Commission, *The Community and Space: A Coherent Approach*, COM (88) 417. Cited in Madders, *A New Force*, p. 571.
67　PE 146.210/Corr. 15 October 1991. Cited in Madders, p. 584.
68　Suzuki, 2003, p. 190.
69　European Commission, *White Paper (COM[2003]673)*, 'Space: a new frontier for an expanding union' (Luxembourg, European Communities, 2003), p. 7.
70　Commission, COM673, p. 7.
71　Suzuki, 2003, p. 197.
72　Suzuki, 2003, p. 195.
73　European Commission, *White Paper on Space*, 2003, p. 9.
74　Michael Sheehan, *International Security: An Analytical Survey* (Boulder, CO, Lynne Rienner, 2005).
75　ESA, *Green Paper on European Space Policy; Report on the Consultation Process* (BR-208) (Noordwijk, Netherlands, ESA Publications Division, 2003), p. 17. The Document called for the EU to develop capabilities in global monitoring, global reconnaissance, monitoring and surveillance including image intelligence and electromagnetic signal analysis, meteorology and oceanography, telecommunications, intelligence information and verification, global command, control, communications and information, global positioning, navigation and timing, mapping and space-based surveillance.
76　ESA, *Green Paper*, 2003, p. 9.
77　*Green Paper*, p. 12.
78　*Green Paper*, p. 15.
79　David Mitrany, 'The functional approach in historical perspective', *International Affairs*, Vol. 47 (1971), p. 532.

6 Space as a military force multiplier

1　Raymond Aron, *Peace and War: A Theory of International Relations* (London, Weidenfeld & Nicolson, 1966), p. 664.
2　Paul B. Stares, *The Militarization of Space: US Policy 1945–1984* (Ithaca, NY, Cornell University Press, 1984), p. 240.
3　Alasdair McLean, *Western European Military Space Policy* (Aldershot, Dartmouth Publishing, 1992), p. 19.

4 Bhupendra Jasani and Christopher Lee, *Countdown to Space War* (London, Taylor & Francis, 1984), pp. 34–6.

5 B. Jasani and T. Sakata, *Satellites for Arms Control and Crisis Monitoring* (Oxford, Oxford University Press, 1987), p. 25.

6 Lt. Gen. Cook, USAF Space Command, quoted in Alasdair McLean, 'A new era? Military space policy enters the mainstream', *Space Policy*, Vol. 16 (2000), p. 244.

7 Curtis Peebles, *Battle for Space* (London, Book Club Associates, 1983), pp. 24–5.

8 Martin Ince, *Space* (London, Sphere Books, 1981), pp. 137–8.

9 Peter Bond, *The Continuing Story of the International Space Station* (Chichester, Springer Praxis, 2002), p. 41.

10 Howell M. Estes III, 'Space and joint space doctrine', *Joint Force Quarterly* (Winter, 1996–7), p. 62.

11 A. Kolovos, 'Why Europe needs space as part of its security and defence policy', *Space Policy*, Vol.18 (2002), P. 260.

12 *Reagan Administration Space Policy Statement*, White House, Washington, DC, 4 July, 1982. Reproduced as Appendix 6 in Hans Mark, *The Space Station: A Personal Journey* (Durham, NC, Duke University Press, 1987), pp. 243–7.

13 Hans Mark, *The Space Station*, pp. 82–3.

14 Alasdair McLean, *Western European Military Space Policy*, p. 19.

15 Peter B. Teets, 'National security space: enabling joint warfighting', *Joint Force Quarterly* (Winter, 2002–3), p. 35.

16 Defense Intelligence Agency, *Soviet Military Space Doctrine* (Washington, DC, DIA, 1984), p. 11.

17 Alasdair McLean, *Western European Military Space Policy*, p. 19.

18 Chairman's Factual Statement, 2002 Preparatory Committee of the Non-Proliferation Treaty, 18 April 2002. Cited in David Grahame, 'A question of intent: missile defence and the weaponisation of space', *Basic Notes*, 1 May 2002, p. 1, available at www.basic.org/pubs/Notes/2002NMDspace.htm, p.1.

19 The White House, National Science and Technology Council, *Fact Sheet: National Space Policy* (Washington, DC, 19 September 1996).

20 Executive Office of the President, *A National Security Strategy for a New Century* (Washington, DC, The White House, December 1999).

21 Thomas D. Bell, 'Weaponisation of space: understanding strategic and technological inevitabilities', *Occasional Paper No. 6*, Centre for Strategy and Technology (Maxwell AFB, AL, United States Air War College, 1999), p. 3.

22 Senator Bob Smith, 'The challenge of space power', *Airpower Journal*, Vol. 13, No. 1 (Spring, 1999), p. 33.

23 Alvin Toffler and Heidi Toffler, *War and Anti-War: Survival at the Dawn of the 21st Century* (New York, Little, Brown & Co., 1993), p. xv.

24 James G. Lee, *Counterspace Operations for Information Dominance* (Maxwell AFB, AL, Air University Press, 1994), p. 19.

25 *Long Range Plan: Implementing USSPACECOM Vision for 2020* (Peterson AFB, CO, US Space Command, March 1998).

26 United States Department of Defense, *News Transcript*, 'Secretary Rumsfeld outlines space initiatives', 8 May 2001.

27 US Space Command, *Long-Range Plan*, p. 7.

28 Everett C. Dolman, *Astropolitik: Classical Geopolitics in the Space Age* (London, Frank Cass, 2002), p. 157.

29 Richard S. Stapp, *Space Dominance: Can the Air Force Control Space?* (Maxwell AFB, AL, USAF Air Command and Staff College, 1997), p. 6.

30 Walter A. McDougall, *The Heavens and the Earth: A Political History of the Space Age* (New York, Basic Books, 1985), p. 194.
31 General Thomas D. White quoted in Richard P. Davenport, *Strategies for Space: Past, Present and Future* (Newport, RI, Naval War College, 1988), p. 1.
32 David N. Spires, *Beyond Horizons: A Half Century of Air Force Space Leadership* (Maxwell AFB, AL; Air University Press, 1998), p. xv.
33 US Congress, House, Select Committee on Astronautics and Space Exploration, *The National Space Programme*, H Rept no. 1758 (85th Congress, 2nd session, 1958), p. 221. Cited in Vernon van Dyke, *Pride and Power: The Rationale of the Space Program* (London, Pall Mall Press, 1965), p. 78.
34 James M. Gavin, *War and Peace in the Space Age* (New York, Harpers, 1958), p. 222.
35 *Baltimore Sun*, 6 July 1980.
36 Curtis Peebles, *The High Frontier; The US Air Force and Military Space Programme* (Washington, DC, Air Force History and Museums Program, 1997), p. 61.
37 Oris B. Johnson, 'Space: today's first line of defense', *Air University Review*, Vol. 20, No. 1 (November–December 1968), p. 96.
38 Paul B. Stares, *The Militarisation of Space, US Policy 1945–1984* (Ithaca, NY, Cornell University Press, 1984), pp. 169–70.
39 Spires, *Beyond Horizons*, p. 192.
40 Paul B. Stares, *The Militarisation of Space*, Chapter 9.
41 David N. Spires, *Beyond Horizons*, pp. 169–71.
42 Max Hastings and Simon Jenkins, *The Battle for the Falklands* (New York, W. W. Norton, 1983).
43 Paul B. Stares, *Space and National Security* (Washington, DC, The Brookings Institution, 1987), p. 46.
44 Thomas A. Keaney and Elliot A. Cohen, *The Gulf War Air Power Summary Report* (Washington, DC, US Government Printing Office, 1993), p. 194.
45 Peter Anson and Dennis Cummings, 'The first space war: the contribution of satellites to the Gulf War', in Alan Cummings (ed.), *The First Information War* (Fairfax, VA, AFCEA International Press, 1992), p. 127.
46 Major Jeffrey Caton, *Joint Warfare and Military Dependence on Space*, available http://fas.org/spp/eprint/lsn3app2.htm, p. 3.
47 Jeffrey L. Caton, 'Joint warfare and military dependence on space', *Joint Force Quarterly* (Winter, 1995–6), p. 49.
48 Howell M. Estes III, 'Space and joint space doctrine', *Joint Force Quarterly* (Winter, 1996–7), p. 62.
49 Cynthia A. S. McKinley, 'When the enemy has our eyes', available http://fas.org/spp/eprint/mckinley.htm, p. 9.
50 R. Cargill Hall and Jacob Neufeld (eds), *The United States Air Force in Space, 1945 to the 21st Century* (Andrews AFB, MD, USAF History and Museums Program, 1995), p. 108.
51 Cargill Hall and Neufeld, p. 118.
52 George M. Moore and Joan Johnson-Freese, 'Joint space doctrine: catapulting into the future', *JFQ* (Summer, 1994), p. 72.
53 General Thomas F. Moorman (Commander of Air Force Space Command during the Gulf War and later Vice Chief of Staff of the Air Force), quoted in Arnold H. Streland, *Clausewitz on Space: Developing Military Space Theory through a Comparative Analysis* (Maxwell AFB, AL, Air Command and Staff College, 1999), pp. 1–2.
54 James G. Lee, *Counterspace Operations for Information Dominance*, available http://fas.org/spp/eprint/lee.htm, p. 1.

55 Robert S. Cooper, 'No. sanctuary: a defense perspective on space', *Issues in Science and Technology*, Vol. 2, No. 3 (Spring, 1986). Reprinted in US Department of Defense, *Earlybird: Special Edition* (Washington, DC, 17 June 1986), p. 19.

56 Steven Lambakis, 'Space control in Desert Storm and beyond', *Orbis* (Summer, 1995), p. 417.

57 Vice-Admiral William Dougherty, 'Storm from space', *Proceedings* (August, 1992), pp. 48–52.

58 M. V. Smith, 'Ten propositions regarding spacepower', *Fairchild Paper* (Maxwell AFB, AL, Air University Press, 2002), p. 25.

59 Donald R. Baucom, 'Space and missile defense', *Joint Force Quarterly* (Winter, 2002–3), p. 51.

60 Donald R. Baucom, p. 52.

61 Ronald D. Humble, *The Soviet Space Programme* (London, Routledge, 1988), p. 78.

62 W. von Kries, 'The demise of the ABM Treaty and the militarization of outer space', *Space Policy*, Vol. 18 (2002), p. 175.

63 B. Jasani, *Space Weapons and International Security* (Oxford, Oxford University Press, 1987), p. 85.

64 J. Pike, 'US and Soviet BMD Programmes', in B. Jasani (ed.), *Outer Space: A Source of Conflict or Cooperation?* (New York, United Nations University Press, 1991), p. 185.

65 http://www.nuclearfiles.org/kmissiledefense/intro.html.

66 Wyn Bowen, 'Missile defence and the transatlantic security relationship', *International Affairs*, Vol. 77, No. 3 (July, 2001), p. 488.

67 B. Jasani, 'US national missile defence and international security: blessing or blight?', *Space Policy*, Vol. 17 (2001), pp. 243–5.

68 Ronald D. Humble, *The Soviet Space Programme* (London, Routledge, 1988), p. 51.

69 Geoffrey E. Perry, 'Russian hunter–killer satellite experiments', *Military Review* (October, 1978), p. 55.

70 Humble, *Soviet Space Programme*, p. 52.

71 Bhupendra Jasani and Christopher Lee, *Countdown to Space War* (London, Stockholm International Peace Research Institute/Taylor & Francis, 1984), pp. 62–3.

72 National Space Policy', NSDD 42, (Washington, DC, Office of the President of the United States, 4 July 1982), available at www.globalsecurity.org/space/library/policy/national/nsdd-42.htm.

73 John Markoff, 'Outer space – the military's new high ground', *Baltimore Sun*, 6 July 1980.

74 Hans Mark, *The Space Station*, pp. 68–69.

75 P. G. Alves (ed.), *Building Confidence in Outer-Space Activities: CSBM's and Earth-to-Space Monitoring* (Aldershot, Dartmouth Publishing, 1996), p. 63.

76 For example, Bell, p. 8.

77 Samuel McNiel, 'Proposed tenets of space power: six enduring truths', *Air and Space Power Journal* (Summer, 2004), p.10 (web offprint).

78 Smith, 'Ten propositions', p. 106.

79 McNiel, 'Proposed tenets of space power', p. 11.

80 McNiel, 'Proposed tenets of space power', p. 11.

81 Gene H. McCall and John A. Corder, *New World Vistas:Air and Space Power for the 21st Century* (Washington, DC, USAF, 1995), p. xiii, available at www.au.af.mil/au/awc/awcgate/vistas/vistas.htm

82 Robert D. Newberry, *Space Doctrine for the 21st Century* (Maxwell AFB, AL, Air University, 1997), p. 50.
83 Colin S. Gray and John Sheldon, 'Space power and the revolution in military affairs: a glass half full?', *Airpower Journal* (Fall, 1999), p. 25.
84 Theresa Hitchens, 'Weapons in space: silver bullet or Russian roulette?', Washington, DC, Centre for Defence Information, 18 April 2002.
85 Peter Hays and Karl Mueller, 'Going boldly – where? Aerospace integration, the Space Commission, and the Air Force's vision for space', *Aerospace Power Journal*, Vol. 15 (Spring, 2001), p. 39.
86 For example, Charles S. Robb, 'Star Wars II', *Washington Quarterly*, 221 (Winter 1999), pp. 81–6.
87 Hays and Mueller, 'Going boldly', p. 39.
88 Bruce M. DeBlois, 'Space sanctuary: a viable national strategy', *Airpower Journal*, Vol. 12, No. 4 (Winter, 1998), p. 41.
89 Bruce M. DeBlois, 'Space sanctuary, p. 57.
90 Hyten, p. 5.
91 Hyten, p. 9.
92 McKinley, 'When the enemy has our eyes', p. 20.
93 John T. Correll, 'Destiny in space', *Air Force Magazine* (August, 1998), p. 2.
94 United States Air Force Doctrine Document, AFDD 2-2.1, *Counterspace Operations* (USAF Doctrine Centre, August, 2004), p. 23.
95 James G. Lee, *Counterspace Operations for Information Dominance*, available http://fas.org/spp/eprint/lee.htm, p. 9.
96 James G. Lee, p. 11.
97 James G. Lee, p. 8.
98 James G. Lee, pp. 4–5.
99 McKinley, p. 28.
100 AFDD 2-2.1, *Counterspace Operations*, p. viii.
101 AFDD 2-2.1, *Counterspace Operations*, p. 28.
102 Patricia Gilmartin, 'Gulf War rekindles US debate on protecting space systems data', *Aviation Week and Space Technology*, Vol. 134, No. 17 (29 April 1991), D.55.
103 Peter B. Teets, 'National security space: enabling joint warfighting', *Joint Force Quarterly* (Winter, 2002–3), p. 37.

7 Space control

1 US President, *Public Papers of the Presidents of the United States, John F Kennedy*, 1 January–31 December 1962 (Washington, DC, US Government Printing Office, 1963), p. 669.
2 United States Air Force Doctrine Document, AFDD 2-2.1, *Counterspace Operations* (USAF Doctrine Centre, August, 2004), p. 54.
3 Howell M. Estes III, 'Space and joint space doctrine', *Joint Forces Quarterly* (Winter, 1996–7), p. 61.
4 General Curtis LeMay, USAF, 1968. Quoted in Richard S. Stapp, *Space Dominance: Can the Air Force Control Space?* (Maxwell AFB, AL, Air Command and Staff College, 1997), p. ?
5 Department of Defense, Joint Publication 1, *Joint Warfare of the Armed Forces of the United States*, 1995, I-3.
6 Richard S. Stapp, *Space Dominance: Can the Air Force Control Space?* (Maxwell AFB, AL, Air Command and Staff College, 1997), p. 9.

7 *Air Force Manual (AFM) 1-2, United States Air Force Basic Doctrine*, 1959; *AFM 1-1, USAF Basic Doctrine*, 1964.
8 *AFM 1-1, United States Air Force Basic Doctrine*, 1971, pp. 2–4.
9 *AFM 1-1, United States Air Force Basic Doctrine*, 1979, pp. 2–8.
10 *AFM 1-6, Military Space Operations*, 1982, p. 1.
11 *AFM 1-6*, p. 5.
12 *AFM 1-6*, pp. 6 and 9.
13 1996 Air Force White Paper, *Global Engagement: A Vision for the 21st Century Air Force*.
14 Defense Intelligence Agency, *Soviet Military Space Doctrine* (Washington, DC, Defense Intelligence Agency, 1984), p. 13.
15 Defense Intelligence Agency, *Soviet Military Space Doctrine* (Washington, DC, Defense Intelligence Agency, 1984), p. 16.
16 V. D. Sokolovski (ed.), *Soviet Military Strategy* (Santa Monica, CA, RAND Corporation, 1963), p. 297. Original published in Moscow, 1962.
17 Ronald D. Humble, *The Soviet Space Programme* (London, Routledge, 1988), p. 35.
18 Defense Intelligence Agency, *Soviet Military Space Doctrine* (Washington, DC, Defense Intelligence Agency, 1984), p. 7.
19 Humble, *The Soviet Space Programme*, p. 46.
20 Defense Intelligence Agency, *Soviet Military Space Doctrine* (Washington, DC, Defense Intelligence Agency, 1984), p. 9.
21 Humble, *The Soviet Space Programme*, p. 46.
22 Curtis Peebles, *Battle for Space* (London, Book Club Associates, 1983), p. 29.
23 Howell M. Estes III, 'Space and joint space doctrine', *Joint Forces Quarterly* (Winter, 1996–7), p. 61.
24 Thomas D. Bell, 'Weaponisation of space: understanding strategic and technological inevitabilities', *Occasional Paper No. 6*, Centre for Strategy and Technology (Maxwell AFB, AL, United States Air War College, 1999), p. 6.
25 Major Jeffrey Caton, *Joint Warfare and Military Dependence on Space*, available at http://fas.org/spp/eprint/lsn3app2.htm, p. 3.
26 'Long range plan: implementing USSPACECOM vision for 2020', March 1998, pp. 19–20.
27 General Lance Lord, Commander, USAF Space Command, 'The argument for space superiority: remarks prepared for Air War College', Maxwell AFB, AL, 29 March 2004, available at www.peterson.af.mil/hqafspc/Library/speeches/Speeches.asp?YearList=2004, p. 7.
28 Lord, p. 8.
29 AFDD 2-2.1, *Counterspace Operations*, p. 55.
30 Thomas A. Torgerson, *Global Power Through Tactical Flexibility: Rapid Deployable Space Units* (Maxwell AFB, AL, Air University Press, 1994), p. 25.
31 George M. Moore, Vic Budura and Joan Johson-Freese, 'Joint space doctrine: catapulting into the future', *Joint Forces Quarterly* (Summer, 1994), p. 75.
32 Klaus Knorr, 'On the international implications of outer space', in *Reflections on Space: Its Implications for Domestic and International Affairs* (Colorado Springs, CO, US Air Force Academy, 1964), p. 247. Cited in Judson J. Jusell, *Space Power Theory: A Rising Star* (Maxwell AFB, AL, Air Command and Staff College, 1998), p. 7.
33 AFDD 2-2.1, *Space Operations* (Maxwell AFB, AL, US Air Force, 1998), p. 1.
34 Colin S. Gray, 'The influence of space power upon history', *Comparative Strategy* (October–December, 1996), p. 293.

35 James E. Oberg, *Space Power Theory* (Colorado Springs, CO, US Air Force Academy, 1999), p. 10.
36 Brent D. Ziarnick, 'The space campaign: space power theory applied to counterspace operations', *Air and Space Power Journal* (Summer, 2004), p. 4.
37 Ziarnick, p. 4.
38 James G. Lee, *Counterspace Operations for Information Dominance*, available at http://fas.org/spp/eprint/lee.htm, p.17.
39 Lee, p. 53.
40 Stapp, *Space Dominance*, p. 14.
41 US General Accounting Office, Report to the Secretary of Defense, *GAO-02–738, Military Space Operations: Planning, Funding, and Acquisition Challenges Facing Efforts to Strengthen Space Control* (Washington, DC, US General Accounting Office, September, 2002), p. 11.
42 1996 Air Force White Paper, *Global Engagement*.
43 Department of the Air Force, AFM 1-1, *Basic Aerospace Doctrine of the United States Air Force*, 1992, p. 6.
44 Todd C. Schull, 'Space operations doctrine: the way ahead', *Air and Space Power Journal* (Summer, 2004), p. 1.
45 AFDD-1, *Air Force Basic Doctrine*, 1997, p. 30.
46 AFDD-1, *Air Force Basic Doctrine*, 2003, p. 52.
47 USAF Scientific Technology Advisory Board, 'Space technology volume', *New World Vistas: Air and Space Power for the 21st Century*, Report to the USAF Chief of Staff (Washington, DC, US Government Printing Office, 1995), p. ix.
48 Samuel McNiel, 'Proposed tenets of space power: six enduring truths', *Air and Space Power Journal* (Summer, 2004), p. 4 (web offprint).
49 AFDD 2-2 (2001), p. 11.
50 US Space Command, *Long-Range Plan: Implementing USSPACECOM Vision for 2020* (Peterson AFB, CO, US Space Command, Director of Plans, 1998), p. 20.
51 US House of Representatives, *Conference Report on S.1059, National Defense Authorisation Act for Fiscal Year 2000*, 106th Congress, 1st session, 5 August 1999, H.R. 106–301, sec. 1621–30, 'Commission to Assess United States National Security Space Management and Organisation'.
52 John E. Hyten, 'A sea peace or a theater of war?', *Air and Space Power Journal* (Fall, 2002), p. 2.
53 *Report of the Commission to Assess United States National Security Space Management and Organization* (Washington, DC, 2001), p. 5.
54 Ibid., p. 29.
55 Ibid., p. 37.
56 Schull, *Space Operations Doctrine*, p. 2.
57 Moore, Budura and Johson-Freese, 'Joint space doctrine', p. 73.
58 Moore, Budura and Johson-Freese., 'Joint space doctrine', p. 76.
59 Schull, *Space Operations Doctrine*, p. 3.
60 Paula Flavell, 'Tenets of air and space power: a space perspective', *Air and Space Power Journal* (Summer, 2004), p. 1. web offprint.
61 Flavell, p. 1.
62 Hon. Peter B. Teets, Under Secretary of the Air Force, 'National security space in the twenty-first century', *Air and Space Power Journal* (Summer, 2004), available at www.airpower.maxwell.af.mil/airchronicles/apj/apj04/sum04/teets.html, p. 1.
63 AFDD 2–2.1, Counterspace operations, *Foreword*.
64 AFDD 2–2.1, p. 1.
65 AFDD 2–2.1, p. 4.

66 Jeffrey L. Caton, 'Joint warfare and military dependence on space', *Joint Forces Quarterly* (Winter, 1995–6), p. 50.
67 AFDD 2-1.1, p. 15.
68 Stapp, *Space Dominance*, p. 14.
69 Bruce M. DeBlois, Richard L. Garwin, R. Scott Kemp and Jeremy C. Marwell, 'Space weapons: crossing the US Rubicon', *International Security*, Vol. 29, No. 2 (Fall, 2004), p. 62.
70 General Howell M. Estes III, 'Space and joint space doctrine', p. 62.
71 Shawn J. Barnes, *Virtual Space Control: A Broader Perspective* (Maxwell AFB, AL, Air Command and Staff College, 1998), p. 1.
72 'Senate committee focuses on military space programs, people', *Air Force News*, 25 March 1999.
73 Hyten, p. 8.
74 Space Commission Report, p. 28.
75 AFDD-1, *Air Force Basic Doctrine*, 1997, pp. 47–8.
76 Barnes, *Virtual Space Control*, p. 8.
77 Michael R. Mantz, *The New Sword: A Theory of Space Combat Power* (Maxwell AFB, AL, Air University Press, 1995), p. 22.
78 James P. Cashin and Jeffrey D. Spencer, *Space and Air Force: Rhetoric or Reality?* (Maxwell AFB, AL, Air Command and Staff College, 1999), p. 35.
79 McKinley, 'When the enemy has our eyes', p. 20.
80 McKinley p. 28.
81 Teets, *National Security Space*, p. 4.
82 General Lance Lord, Commander, USAF Space Command, 'Remarks prepared for Air War College', Maxwell AFB, AL, 29 March 2004, available at www.peterson.af.mil/hqafspc/Library/speeches/Speeches.asp?YearList=2004.
83 Bruce M. DeBlois, 'Space sanctuary: a viable national strategy', *Aerospace Power Journal* (Winter, 1998).
84 *Congressional Record*, Vol. 104, pt 8 (2 June 1958), 9913.
85 US Congress, Senate, Committee on Aeronautical and Space Sciences, *Documents on International Aspects of the Exploration and Use of Outer Space, 1954–1962*, Staff report, S.Doc, no. 18 (88th Congress, 1st session, 1963), p. 253.
86 President of the United States, *US National Space Policy* (Washington, DC, White House, 1996).
87 The White House, *A National Security Strategy for a New Century*, May, 1997.
88 General Lance Lord, p. 8.
89 AFDD 2-2.1, *Counterspace Operations*, p. 39.
90 Moore, Budura and Johnson-Freese, 'Joint space doctrine', p. 75.
91 Henry G. Franke III, *An Evolving Joint Space Campaign Concept and the Army's Role* (Fort Leavenworth, KS, Army Command and General Staff College, 1992), p. 19.
92 Erik Berghaust, *The Next Fifty Years in Space* (New York, Macmillan, 1964), p. 101.
93 Barnes, *Virtual Space Control*, p. 36.
94 *Statement by John R Bolton, US Under Secretary of State for Arms Control and International Security*, to the Conference on Disarmament, Geneva, 24 January, 2002. Cited in David Grahame, 'A question of intent: missile defence and the weaponisation of space', *Basic Notes*, 1 May 2002, p. 1, available at www.basic.org/pubs/Notes/2002NMDspace.htm, p.2.
95 Cashin and Spencer, *Space and Air Force*, p. 24.

8 Space, justice and international development

1 Anna K. Dickson, *Development and International Relations: A Critical Introduction* (Cambridge, Polity Press, 1997), p. 19.

2 Joan Spero, *The Politics of International Economic Relations* (London, Unwin Hymen, 1985), pp. 169–70.

3 *Working Paper on Principles Regarding International Cooperation in the Exploration and Utilisation of Outer Space for Peaceful Purposes.* UN Doc. A/AC.105/C.2/L.182/Rev.1. (New York, United Nations Organization, 31 March 1993).

4 Hubert George, 'Remote sensing of Earth resources: emerging opportunities for developing countries', *Space Policy*, Vol. 14 (1998), p. 27.

5 S. E. Doyle *et al.*, 'Legal practicalities and realities of the use of outer space by developing countries', *Proceedings of the 32nd Colloquium of the Law of Outer Space* (1989), pp. 3–4.

6 M. S. Soroos, 'The commons in the sky: the radio spectrum and geosynchronous orbit as issues in global policy', *International Organisation*, Vol. 36 (1982), p. 668.

7 E. Miles, 'Transnationalism in space: inner and outer', *International Organisation*, Vol. 25 (1971), p. 625.

8 R. Chipman, *The World in Space* (Englewood Cliffs, NJ, Prentice Hall, 1982), pp. 504–14.

9 OTA, *International Cooperation and Competition in Civilian Space Activities* (Washington, DC, US Congress, Office of Technology Assessment, OTA-ISC-239, July 1985), p. 326.

10 Hubert George, 'Remote sensing of Earth resources', p. 28.

11 Hubert George, 'Remote sensing of Earth resources', p. 28.

12 Arnold Frutkin, 'International cooperation in space', *Science*, 24 July 1970.

13 Hubert George, 'Remote sensing of Earth resources', p. 30.

14 M. L. Smith, 'Equitable access to the orbit/spectrum resource', in *Proceedings of the 30th Colloquium of the Law of Outer Space* (1988), p. 264.

15 Peter Hulsroj, 'Beyond global: the international imperative of space', *Space Policy*, Vol. 18 (2002), p. 107.

16 C. Specter, 'ISY: an opportunity to relate earth observation activities to the benefits and interests of the developing countries', *Proceedings of the 32nd Colloquium on the Law of Outer Space* (1990).

17 *Treaty on Principles Governing the Activities of States in the Exploration and Use of Outer Space, Including the Moon and Other Celestial Bodies, 1967.*

18 N. Jasentuliyana, 'Review of the recent discussions relating to aspects of article one of the Outer Space Treaty', in *Proceedings of the 29th Colloquium of the Law of Outer Space* (1987).

19 Prityatna Abdurrasyid, 'Developing countries and the use of the geo-stationary orbit', *Proceedings of the 30th Colloquium of the Law of Outer Space*, p. 375.

20 Joseph N. Pelton, 'The proliferation of communications satellites: gold rush in the Clarke Orbit', in James E. Katz, *People in Space* (Oxford, Transaction Books, 1985), p. 105.

21 Edward Miles, 'Transnationalism in space: inner and outer', *International Organisation*, Vol. 25, No. 3 (Summer, 1971), p. 604.

22 Peter Hulsroj, 'Beyond global: the international imperative of space', *Space Policy*, Vol. 18 (2002), p. 108.

23 Siegfried Weissner, 'Access to a *Res Publica Internationalis*', in *Proceedings of the 29th Coloquium of the Law of Space* (1987).

24 Michael Laver, 'The politics of inner space: tragedies of three commons', *European Journal of Political Research*, Vol. 12 (1984), p. 65.

25 *1973 ITU Convention* (also known as the Malaga–Torremolinos Convention), Article 33.

26 Joseph N. Pelton, 'The proliferation of communications satellites', p. 103.

27 Marvin S. Soroos, 'The commons in the sky', p. 377.

28 Anna K. Dickson, *Development and International Relations*, p. 4.

29 Bin Cheng, 'The legal regime of air space and outer space: the boundary problem', *Annals of Air and Space Law*, Vol. 5 (1980), p. 323ff.

30 Ram S. Jakhu, 'The legal status of the geostationary orbit', *Annals of Air and Space Law*, Vol. 7 (1982), p. 341.

31 United Nations Organization, *Agreement Governing the Activities of States on the Moon and Other Celestial Bodies*, New York, 18 December 1979, available at www.greaterearth.org/laws/moon_try.htm.

32 S. Weissner, 'Access to a *Res Publica Internationalis*', p. 149.

33 Radio Regulations Article S23.13, and Appendix 30.

34 OTA, *International Cooperation*, p. 175.

35 Michael Laver, 'The politics of inner space', p. 62.

36 Soroos, 1982, pp. 666–72.

37 N. Jasentuliyana, 'Ensuring equal access to the benefits of space technologies for all countries', *Space Policy*, Vol. 10 (1994), p. 11.

38 Peter Hulsroj, 'Beyond global', p. 115.

39 D. Demac, *Tracing New Orbits* (New York, Columbia University Press,1986), p. 251.

40 OTA, *International Cooperation*, p. 176.

41 Resolution No. 3, 1979 WARC. Cited in Demac (1986), p. 209.

42 UNISPACE 82 Report. Cited in Y. S. Rajan, 'Benefits from space technology: a view from a developing country', *Space Policy*, Vol. 4 (1988), p. 222.

43 *Report of the Second United Nations Conference on the Exploration and Peaceful Uses of Outer Space*, Vienna, 9–21 August 1982, UN Document, A/CONF.101/10.

44 Demac (1986), pp. 32–9.

45 Hubert George, 'Developing countries and remote sensing: how intergovernmental factors impede progress', *Space Policy*, Vol. 16 (2000), p. 268

46 OTA, *International Cooperation*, p. 329.

47 Demac (1986), pp. 58–75.

48 Sylvia Maureen Williams, 'Direct broadcast satellites and international law', *International Relations*, Vol. 8 (1985), p. 245.

49 Economic Commission for Africa, *African Information Society Initiative* (1996), p. 59.

50 Marietta Benko and Kai-Uwe Schrogl, 'Space benefits: towards a useful framework for international cooperation', *Space Policy*, Vol. 11 (1995), p. 5.

51 UN General Assembly Resolution 51/122, 13 December 1996, *Declaration on International Cooperation in the Exploration and Use of Outer Space for the Benefit and in the Interest of All States, Taking into Particular Account the Needs of the Developing Countries*.

52 UNGA Resolution 51/122.

53 Juan G. Roederer, 'The participation of developing countries in space research', *Space Policy*, Vol. 1 (1985), p. 316.

54 Benko and Schrogl, 'Space benefits', p. 5.

55 Nigel Dower, 'The idea of international development: some ethical issues', paper presented at the European Consortium for Political Research, Madrid,

1994. Quoted in Anna K. Dickson, *Development and International Relations: A Critical Introduction* (Cambridge, Polity Press, 1997), p. 22.

9 India: security through space

1　Brian Harvey, *The Japanese and Indian Space Programmes: Two Roads into Space* (Chichester, Springer Praxis Books, 2000), p. 127.
2　Nicholas Nugent, 'The defence preparedness of India: arming for tomorrow', *Military Technology* (March, 1991), p 32.
3　ISRO, *Space India*, web edition, available at www.isro.org/newsletters/spaceindia/oct2004jun2005/Chapter3.htm.
4　For example, S. K. Gosh, 'India's space programme and its military implications', *Asian Defence Journal* (September, 1981); Raju G. C. Thomas, 'India's nuclear and space programmes: defence or development?', *World Politics*, Vol. 38 (1985–6), pp. 315–42; Anita Bhatia, 'India's space programme: cause for concern?', *Asian Survey* (October, 1985), pp. 1013–30; Thomas G. Mahnken, 'Why Third World space systems matter', *Orbis*, Vol. 35 (1991), pp. 571–2.
5　Michael Sheehan, 'The evolution of the concept of security since 1945', in Michael Sheehan (ed.), *National and International Security* (Aldershot, Ashgate, 2000), pp. 1–31; Michael Sheehan, *International Security: An Analytical Survey* (Boulder, CO, Lynne Rienner, 2005).
6　www.isro.org/citizencharter.htm.
7　www.prl.res.in/~simba/rafiki/space_prog.html.
8　Indian Space Research Organisation, *Department of Space, Annual Report, 2005–6*.
9　Mohan Sundara Rajan, *India in Space* (New Delhi, Government of India, Ministry of Information and Broadcasting, 1979), preface. See also Government of India, Department of Space, *Annual Report 1995–96* (New Delhi, 1996), p. 109.
10　Stephan F. von Welck, 'India's space policy: a developing country in the space club', *Space Policy*, Vol. 3 (1987), p. 330.
11　Kapil Kaul, 'India in space', *Strategic Analysis* (New Delhi), Vol. 10 (1986), p. 854.
12　von Welck, p. 326.
13　Ray Harris, 'Current policy issues in remote sensing: report by the International Policy Advisory Committee of ISPRS', *Space Policy*, Vol. 19 (2003), p. 295.
14　Brian Harvey, *The Japanese and Indian Space Programmes*, p. 127.
15　Aaron Karp, 'Ballistic missiles in the Third World', *International Security*, Vol. 9 (1984–5), p. 180.
16　Mahnken, pp. 563–4.
17　Y. S. Rajan, 'Benefits from space technology: a view from a developing country', *Space Policy*, Vol. 4 (1988), p. 225.
18　M. S. Rajan, *India in Space*, p. 81.
19　*Report of the Second United Nations Conference on the Exploration and Peaceful Uses of Outer Space*, Vienna, 9–21 August 1982. UN. A/Conf.101/10; para. 11.
20　Brian Harvey, *The Japanese and Indian Space Programmes*, p. 130.
21　*Jane's Spaceflight Directory* (London, Jane's, 1987), p. 31.
22　Janne E. Nolan, *Trappings of Power: Ballistic Missiles in the Third World* (Washington, DC, Brookings Institution, 1991), p. 42.
23　Jerrold Elkins and Brian Fredericks, 'Military implications of India's space program', *Air University Review*, Vol. 34 (1983), p. 57.
24　Elkins and Fredericks, p. 58.

25 Mohan S. Rajan, p. 45.
26 Brian Harvey, *The Japanese and Indian Space Programmes*, p. 144.
27 Ray Harris, 'Current policy issues in remote sensing: report by the International Policy Advisory Committee of ISPRS', *Space Policy*, Vol. 19 (2003), p. 295.
28 Government of India, Department of Space, *Annual Report 1995–96*, p. 34.
29 S. Bhatt, 'Some perspectives on outer space exploration by India', *Indian Quarterly*, Vol. 32 (1976), p. 23.
30 Brian Harvey, *The Japanese and Indian Space Programmes*, p. 133.
31 Mohan Sundara Rajan, *India in Space*, p. 52.
32 U. R. Rao (Chairman, Space Commission), 'Space technology in developing nations: an assessment', *Space Policy* (1993), p. 167.
33 *Space News*, 18–24 June 1990.
34 Brian Harvey, *The Japanese and Indian Space Programmes*, pp. 146, 151.
35 Anil Ananthaswamy, 'India special: space programme presses ahead', New Scientist.com news service, 19 February 2005, available at www.newscientist. com/special/india/mg18524871.000.
36 D. A. Brown, 'India developing INSAT follow-on to serve 1990s national needs', *Aviation Week and Space Technology* (12 May 1986), p. 56.
37 Mohan Sundara Rajan, *India in Space*, p. 67.
38 Brian Harvey, *The Japanese and Indian Space Programmes*, p. 142. NNRMS programmes are initiated and coordinated through standing committees on agriculture and soils, bio-resources, geological resources, water resources, ocean resources, cartography and mapping, rural development, technology and training and meteorology.
39 Brian Harvey, *The Japanese and Indian Space Programmes*, p. 140.
40 *Space News* (26 August–8 September 1991), p. 22.
41 Indian Space Research Organisation, Department of Space, available at www. isro.org/rep2002/Links/Earth per cent20.htm.
42 Anil Ananthaswamy, 'India special: space programme presses ahead', New Scientist.com news service, 19 February 2005, available at www.newscientist. com/special/india/mg18524871.000.
43 Brian Harvey, *The Japanese and Indian Space Programmes*, p. 143.
44 Anil Ananthaswamy, 'India special: Space programme presses ahead', New Scientist.com news service, 19 February 2005, available at www.newscientist. com/special/india/mg18524871.000.
45 Ray Harris, 'Current policy issues in remote sensing', pp. 294–5.
46 www.isro.org/rep2002/Links/Organisation.htm.
47 von Welck, 'India's space policy', p. 331.
48 Indian Space Research Organisation, 2003, available at www.isro.org/rep2002/ Links/Earth per cent20.htm.
49 Indian Space Research Organisation, 2003, available at www.isro.org/rep2002/ Links/Earth per cent20.htm.
50 Bhatt, 'Some perspectives', p. 20.
51 For example, S. K. Gosh, 'India's space programme'; Raju G. C. Thomas, 'India's nuclear and space programmes', pp. 315–42; Bhatia, 'India's space programme', pp. 1013–30; Mahnken, pp. 571–2.
52 H. P. Mama, 'India's space programme: across the board on a shoestring', *Interavia*, Vol. 35 (January, 1980), p. 60.
53 H. P. Mama, p. 60.
54 Bhatia, 'India's space programme', pp. 1023, 1029.
55 Federation of American Scientists, available at www.fas.org/spp/guide/india/ earth/irs.html.

56 Mahnken, pp. 571–2.
57 www.flonnet.com/fl1823/18230780.htm.
58 Brian Harvey, *The Japanese and Indian Space Programmes*, p. 161.
59 C. Raju Mohan and K. Subrahmanyam, 'High technology weapons in the developing world', in Eric H Arnott (ed.), *New Technologies for Security and Arms Control: Threats and Promises* (Washington, DC, American Association for the Advancement of Science, 1989), p. 229.
60 J. Mohan Malik, 'India copes with Kremlin's fall', *Orbis*, Vol. 37 (1993), p. 84.
61 Nugent, 'The defence preparedness of India', p. 31.
62 *The Hindu* (Madras), 15 August, 1980.
63 'Zero option for India', *Vikrant* (November, 1981), p. 2, quoted in Elkin and Fredericks, p. 59.
64 Aaron Karp, 'Ballistic missile proliferation', in *SIPRI Yearbook 1991: World Armaments and Disarmament* (Oxford, Oxford University Press, 1991), p. 324.
65 Bhatia, 'India's space programme', p. 178.
66 www.fas.org/nuke/guide/india/missile/agni-htm.
67 M. Navias, 'Ballistic missile proliferation in the Third World', *Adelphi Paper 252* (London, International Institute for Strategic Studies, 1990), p. 13.
68 J. Mohan Malik, p. 83.
69 G. S. Bhargava, 'India's security in the 1980's', *Adelphi Paper No. 125* (London, International Institute for Strategic Studies, 1976), p. 14.
70 Raju G. C. Thomas, p. 324.
71 Jonas Bernstein, 'Beijing nuclear strategy makes China a major player', *Insight* (22 August 1988), p. 30.
72 Jasjit Singh, 'Indian security: a framework for national strategy', *Strategic Analysis* (New Delhi, November, 1987), p. 898.
73 www.globalsecurity.org/org/news/2001/010419–gslv.htm.
74 www.isro.org/rep2002/links/Space per cent20transportation.htm
75 G. Milholin, *Research Report: West German Aid to India's Rocket Programme* (Washington, DC, Wisconsin Project on Nuclear Arms Control, April, 1989).
76 Nolan, p. 45.
77 J. Mohan Malik, p. 86.
78 Barry Buzan, Ole Waever and Jaap de Wilde, *Security: A New Framework for Analysis* (Boulder, CO, Lynne Rienner, 1998).
79 von Welck, p. 327.
80 von Welck, p. 332.
81 S. Bhatt, 'Some perspectives on outer space exploration by India', *Indian Quarterly*, Vol. 32 (1976), p. 20.
82 Brian Harvey, *The Japanese and Indian Space Programmes*, p. 178.
83 von Welck, p. 327; Mohan Rajan, pp. 81–2.
84 K. S. Jayaraman, 'India's space agency proposes manned spaceflight program', *Space News*, 10 November 2006, available at www.space.com/news/061110_india_mannedspace.htm.
85 K. S. Jayaraman, 'India's space agency proposes manned spaceflight program', *Space News*, 10 November 2006, available at www.space.com/news/061110_india_mannedspace.htm.
86 'India mulls human space mission', *New Scientist*, 6 November 2006, available at http://space.newscientist.com/article/dn10458.

10 China: the long march into space

1 William R. Morris, 'The role of China's space program in its national development strategy', *Maxwell Paper No. 24* (Maxwell AFB, AL, Air War College, August 2001), p. 3.
2 Jeff Kingswell, 'The rise of the Asian space dragon', *Space Policy*, Vol. 10 (1994), p. 188.
3 Rosita Dellios, 'China's space programme: a strategic and political analysis', *Culture Mandala*, Vol. 7, No. 1 (December, 2005), available at www.international-relations.com/CM7–1Wb/ChinasSpaceWB.htm, p.2.
4 James A. Lewis, 'China as a military space competitor' (Washington, DC, Centre for Strategic and International Studies, August 2004), p. 1.
5 Dellios, 'China's space programme: a strategic and political analysis', p. 1.
6 Dellios, p. 1.
7 Michael Sheehan, 'Chinese military space policy', *Enjeux Atlantiques*, Vol. 4 (1991), p. 43.
8 Yanping Chen, 'China's Space policy: a historical review', *Space Policy*, Vol. 7 (1991), p. 119.
9 Yanping Chen, 'China's space policy', 119.
10 P. S. Clark, 'The Chinese space programme', *Journal of the British Interplanetary Society*, Vol. 37 (1984), p. 195.
11 Moscow Radio, 16 April 1968. Cited in Dellios, p. 3.
12 Yanping Chen, 'China's space policy', p. 122.
13 Yanping Chen, 'China's space policy', p. 118.
14 Yanping Chen, 'China's space policy', p. 120.
15 Anne Gilks, 'China's space policy: review and prospects', *Space Policy*, Vol. 13 (1997), p. 215.
16 Brian Harvey, 'China's space programme: emerging competitor or potential partner?', Centre for Nonproliferation *Occasional Paper, No. 12*, p. 50.
17 Yanping Chen, 'China's space policy', p. 120.
18 John Wilson Lewis and Xue Litai, *China's Strategic Seapower: The Politics of Force Modernisation in the Nuclear Age* (Stanford, CA, Stanford University Press, 1994), p. 165.
19 Joan Johnson-Freese and Andrew S. Erickson, 'The emerging China–EU space partnership: a geotechnological balancer', *Space Policy*, Vol. 22 (2006), p. 13.
20 P. S. Clark, 'The Chinese space programme', p. 200.
21 Brian Harvey, 'China's space programme', p. 51.
22 Yanping Chen, 'China's space policy: a historical review', p. 124.
23 Kenneth G. Weiss, 'Space dragon: long march, proliferation and sanctions', *Comparative Strategy*, Vol. 18 (1999), p. 347.
24 Chen Lan, 'Hard road to commercial space', *Spaceflight*, Vol. 48, No. 11 (November, 2006), p. 428.
25 Chen, 'China's space policy: a historical review', p. 126.
26 Johnson-Freese and Erickson, p. 13.
27 Johnson-Freese and Erickson, p. 13.
28 Johnson-Freese and Erickson, p. 13.
29 Yanping Chen, *Space Policy*, Vol. 7 (1991), p. 126.
30 Office of the US Secretary of Defense, *Annual Report to Congress: The Military Power of the People's Republic of China*, available at http://www.globalsecurity.org/military/library/report/2004/d20040528prc.pdf; and www.defenselink.mil/news/Jul2005/d20050719china.pdf Both cited in Dellios, p. 6.
31 Chen Lan, 'Hard road to commercial space', *Spaceflight*, Vol. 48, No. 11 (November, 2006), p. 428.

32 Chen Lan, 'Hard road to commercial space', p. 430.
33 Zhuang Fengan, Director, China Aerospace Corporation Science and Technology Committee, cited in, William C. Martel and Toshi Yoshihara, 'Averting a Sino–US space race', *Washington Quarterly*, Vol. 26, No. 4 (Autumn, 2003), pp. 25–6.
34 Phillip C. Saunders, 'China's future in space: implications for US security', *adAstra: The Magazine of the National Space Society*, available at http://www. space.com/adastra/China_implications_0505.html, 20 September 2005.
35 *Report of the Commission to Assess United States National Security Space Management and Organization* (Washington, DC, USGPO, 11 January, 2001), . www.space.gov/docs/fullreport.pdf.
36 William R. Morris, p.12.
37 Qiu Yong, 'Analysis on the ASAT capability of the GMD interceptor', Presentation at the 16th International Summer Symposium on Science and World Affairs, Beijing, 17–25 July 2004. Cited in Hui Zhang, 'Action–reaction: US space weaponisation and China', *Arms Control Today* (December, 2005), available at http://www.armscontrol.org/act/2005_12?Dec-cvr.asp?print, p.1.
38 *Report of the Commission to Assess United States National Security Space Management and Organization* (Washington, DC, US Government Printing Office, 2001).
39 AFDD 2-2.1, Counterspace operations (Colorado Springs, CO, Air Force Doctrine Centre, August 2004).
40 *US National Space Policy*, White House, 2006, available at www.ostp.gov/html/ US%20National%20Space%20Policy.pdf.
41 John W. Garver, 'China's response to the SDI', *Asian Survey* (November, 1986), p. 1221. See also, Bonnie S. Glaser and Banning N. Garrett, 'Chinese perspectives on the SDI', *Problems of Communism* (March–April, 1986).
42 Michael Swaine, 'Assessing the meaning of the Chinese ASAT test', available at www.carnegieendowment.org/publications/index.cfm?fa=view&id=19006&p rog=zch.
43 Rosita Dellios, 'China's space programme', p.8.
44 Jeffrey Lewis, 'Engage China, engage the world', *adAstra* (May, 2005), available at www.space.com/adastra/china_engagement_0505.html.
45 William C. Martel and Toshi Yoshihara, 'Averting a Sino–US space race', *Washington Quarterly*, Vol. 26, No. 4 (Autumn, 2003), p. 31.
46 Joan Johnson-Freeze, 'Scorpions in a bottle: China and the US in space', *Non-Proliferation Review*, Vol. 11, No. 2 (Summer, 2004), p. 179.
47 'Russia and China's moon pact', *Spaceflight*, Vol. 48, No. 11 (November, 2006), p. 405.
48 CNN.com, 'Bush offers China space cooperation', available at http://edition. cnn.com/2006/TECH/space/04/21/China.us.space.reut/index.html, 21 April 2006.
49 Johnson-Freese and Erickson, p. 14.
50 Gilks, p. 216.
51 James A. Lewis, 'China as a military space competitor' (Washington, DC, Centre for Strategic and International Studies, August 2004), p. 2.
52 Wu Guoxiang, *Space Policy*, Vol. 4 (1988), pp. 41–2; Zhu Yilin, *Space Policy*, Vol. 9 (1993), pp. 171–2. Cited in Gilks, p. 217.
53 David J. Thompson, 'China's military space programme: strategic threat, regional power, or national defense?', *Maxwell Paper No. 24* (Maxwell AFB, AL, Air War College, August 2001), p. 21.
54 Gilks, p. 218.
55 *Space News*, 15–28 November 1993. Cited in Gilks, p. 219.

56 Gilks, p. 219.
57 Gilks, p. 216.
58 Michael Sheehan, 'Mars calling: China's space programme', *The World Today* (November, 2003), p. 17.
59 Brian Harvey, 'China's space programme', p. 51.
60 Rosita Dellios, 'China's space programme' p. 4.
61 Brian Harvey, 'China's space programme', p. 51.
62 Brian Harvey, 'China's space programme', p. 51.
63 Yanping Chen, *Space Policy*, Vol. 7 (1991), p. 124.
64 James A. Lewis, 'China as a Military Space Competitor' (Washington, DC, Centre for Strategic and International Studies, August 2004), p. 1.
65 Rosita Dellios, 'China's space programme', p. 5.
66 Rosita Dellios, 'China's space programme', p. 5.
67 'China's space programme aims at peaceful use of space resources', Chinanews. cn, 15 October 2005, available at www.chinanews.cn/news/2005/2005–10–15/12428.html.
68 China National Space Administration. Press Release, 30/12/2003, avaulable at www.cnsa.gov.cn/english/news_release/show.asp?id=89.
69 Zhu Yilin and Xu Fuxiang, 'Status and prospects of China's space programme', *Space Policy*, Vol. 13 (1997), p. 70.
70 Zhu Yilin and Xu Fuxiang, p. 70.
71 Hu Shixiang, Deputy Director of China's manned spaceflight programme, quoted in Min Lee, 'China aims to put man on the moon by 2020', Space.com, 27 November, 2005, available at www.space.com/missionlaunches/ap_051127_china_moon.html.
72 Johnson-Freese and Erickson, p. 13.
73 Rosita Dellios, 'China's space programme', p. 6.
74 Martel and Yoshihara, p. 31.
75 Rosita Dellios, 'China's space programme: a strategic and political analysis', *Culture Mandala*, Vol. 7, No. 1 (December, 2005), available at www.international-relations.com/CM7–1Wb/ChinasSpaceWB.htm p. 5.

11 Cooperation and competition in the post-Cold War era

 1 OTA, *International Cooperation and Competition in Civilian Space Activities* (Washington, DC, US Congress, Office of Technology Assessment, OTA-ISC-239, July, 1985), p. 373.
 2 Andrew Smith, *Moondust* (London, Bloomsbury, 2005), p. 300.
 3 Unclassified version of Presidential Directive-42, White House, 10 October, 1978. Reproduced in Hans Mark, *The Space Station: A Personal Journey* (Durham, NC, Duke University Press, 1987), p. 231.
 4 Hans Mark, *The Space Station: A Personal Journey* (Durham, NC, Duke University Press, 1987), p. 108.
 5 Howard E. McCurdy, *The Space Station Decision: Incremental Politics and Technological Choice* (Baltimore, MD, Johns Hopkins University Press, 1990), p. 177.
 6 McCurdy, *The Space Station Decision*, p. 40.
 7 McCurdy, *The Space Station Decision*, p. 177.
 8 W. H. Lambright, 'Leadership and large-scale technology: the case of the international space station', *Space Policy*, Vol. 21 (2005), p. 196.
 9 John Logsdon, *Together in Orbit: The Origins of International Participation in the Space Station* (Washington, DC, NASA, 1998).

10 Hans Mark, *The Space Station*, pp. 180–1.
11 President Ronald Reagan, *State of the Union Address* (Washington, DC, 1984), available at http://www.presidentreagan.info/speeches/reagan_sotu_1984.cfm.
12 John Logsdon, 'Foreign policy in orbit: the international space station', available at www.afsa.org/fsj/apr01/logsdonapr01.cfm.
13 David M. Harland and John E. Catchpole, *Creating the International Space Station* (Guildford, Springer-Verlag, 2002), p. 121.
14 Peter Bond, *The Continuing Story of the International Space Station* (Chichester, Springer Praxis, 2002), p. 52.
15 Eligar Sadeh, 'Technical, organisational and political dynamics of the international space station', *Space Policy*, Vol. 20 (2004), p. 173.
16 Matthew von Bencke, *The Politics of Space: A History of US–Soviet/Russian Cooperation and Competition in Space* (Boulder, CO, Westview Press), p. 98.
17 Harland and Catchpole, *Creating the International Space Station*, pp. 125–6.
18 John Logsdon, *Together in Orbit*, p. 22.
19 John Logsdon, *Together in Orbit*, p. 26.
20 http://eisenhowerinstitute.org/programs/globalpartnerships/fos/newfrontier/The%20International%20Space%20Station%20–%20The %20View%20from%20Japan%20–%20Kori%20Urayama%20and%20Shuichi%20Wada.pdf.
21 Piers Bizony, *Island in the Sky: Building the International Space Station* (London, Aurum Press, 1996), p. 47.
22 W. D. Kay, *Defining NASA: The Historical Debate Over the Agency's Mission'* (Albany, NY, State University of New York Press, 2005), p. 157.
23 *Agreement between the United States of America and the Russian Federation Concerning Cooperation in the Exploration and Use of Outer Space for Peaceful Purposes*, available at http://www.jaxa.jp/jda/library/space-law/chapter_4/4–2–2–6_e.html.
24 Susan Eisenhower, *Partners in Space: US–Russian Cooperation after the Cold War* (Washington, DC, The Eisenhower Institute, 2005), p. VIII.
25 Peter Bond, *The Continuing Story of the International Space Station*, p. 126.
26 Yuri Karash, *The Superpower Odyssey: A Russian Perspective on Space Cooperation* (Reston, VA, American Institute of Aeronautics and Astronautics,1999), p. 180.
27 John M. Logsdon and James R. Millar, *US–Russian Cooperation in Human Space Flight; Assessing the Impacts*, available at www.gwu.edu/~spi/usrussia.html.
28 John Logsdon, 'Foreign policy in orbit: the international space station', available at www.afsa.org/fsj/apr01/logsdonapr01.cfm.
29 John Pike, *Ralpha: Russian American Space Cooperation*, available at http://www.fas.org/spp/eprint/jp_931210.htm.
30 Goldin quoted in Yuri Karash, *The Superpower Odyssey*, p. 190.
31 Brian Harvey, *Europe's Space Programme: To Ariane and Beyond* (Chichester, Springer Praxis, 2003), p. 314.
32 Peter Bond, *The Continuing Story*, p. 106.
33 Alexander Yakovenko, 'The intergovernmental agreement on the international space station', *Space Policy*, Vol. 15 (1999), pp. 79–86.
34 Sharon Squassoni and Marcia S. Smith, *The Iran Non-proliferation Act and the International Space Station: Issues and Options*. Congressional Research Service (Washington, DC,), Available at http://www.fas.org/sgp/crs/space/RS22072.pdf
35 Brian Berger, 'Senate clears NASA to buy Russian spaceships', Space.com, available at www.space.com/news/050921_senate_soyuz.html.
36 Reimar Lust, *Die Zeit*, 10 March 1995.
37 Susan Eisenhower, *Partners in Space*, p. X.

38 Steven Berner, *Japan's Space Program: A Fork in the Road?* (Santa Monica, CA, RAND Corporation, 2005), p. 21.
39 'China and Japan launch race to the Moon', http://www.guardian.co.uk/space/ article/0,14493,1441062,oo.html
40 Mark Sappenfield, 'India raises the ante on its space program'. *Christian Science Monitor*, 17 January 2007.
41 *A Renewed Spirit of Discovery* (Washington, DC, The White House, 2004). http://www.whitehouse.gov/infocus/space/

Bibliography

Abdurrasyid, Prityatna, 'Developing countries and the use of the geo-stationary orbit', *Proceedings of the Thirtieth Colloquium of the Law of Outer Space* (Reston, VA, American Institute of Aeronautics and Astronautics, 1988).

Air Force Doctrine Center, *AFDD-2, Space Operations* (Colorado Springs, CO, United States Air Force, 1998).

Air Force Manual (AFM) 1–2, United States Air Force Basic Doctrine, 1959; *AFM 1–1, USAF Basic Doctrine* (Washington, DC, Department of the Air Force, 1964).

Air Force White Paper, *Global Engagement: A Vision for the 21st Century Air Force* (Washington, DC, Department of the Air Force, 1996).

Aldridge, E. C. Jnr, 'The myths of militarization of space', *International Security*, Vol. 11 (1987), pp. 151–6.

Aldrin, Buzz and McConnell, Malcolm, *Men from Earth* (London/New York, Bantam Press, 1989).

Alves, P. G. (ed.), *Building Confidence in Outer-Space Activities: CSBMs and Earth-to-Space Monitoring* (Aldershot, Dartmouth Publishing, 1996).

Ananthaswamy, Anil, 'India special: space programme presses ahead', New Scientist. com news service, 19 February 2005, available at www.newscientist.com/special/ india/mg18524871.000.

Anson, Peter and Cummings, Dennis, 'The first space war: the contribution of satellites to the Gulf War', in Alan Cummings (ed.), *The First Information War* (Fairfax, VA, AFCEA International Press, 1992).

Aron, Raymond, *Peace and War: A Theory of International Relations* (London, Weidenfeld & Nicolson, 1966).

Barghoorn, F. C., *Soviet Foreign Propaganda* (Princeton, NJ, Princeton University Press, 1964).

Barnes, Shawn J., *Virtual Space Control: A Broader Perspective* (Maxwell AFB, AL, Air Command and Staff College, 1998).

Baucom, Donald R., 'Space and missile defense', *Joint Force Quarterly* (Winter, 2002–3), pp. 50–5.

Bell, Thomas D., 'Weaponisation of space: understanding strategic and technological inevitabilities', *Occasional Paper No. 6*, Center for Strategy and Technology (Maxwell AFB, AL, United States Air War College, 1999).

Benko, Marietta and Schrogl, Kai-Uwe, 'Space benefits – towards a useful framework for international cooperation', *Space Policy*, Vol. 11 (1995), pp. 5–8.

Berghaust, Erik, *The Next Fifty Years in Space* (New York, Macmillan, 1964).

Bernstein, Jonas, 'Beijing nuclear strategy makes China a major player', *Insight* (22 August 1988).

Beschlass, Michael R., 'Kennedy and the decision to go to the moon', in Roger D. Launius and Howard E. McCurdy (eds), *Spaceflight and the Myth of Presidential Leadership* (Champaign, IL, University of Illinois Press, 1997).

Bhargava, G. S., 'India's security in the 1980s', *Adelphi Paper No. 125* (London, International Institute for Strategic Studies, 1976).

Bhatia, Anita, 'India's space programme: cause for concern?', *Asian Survey* (October, 1985), pp. 1013–30.

Bhatt, S., 'Some perspectives on outer space exploration by India', *Indian Quarterly*, Vol. 32 (1976), pp. 18–25.

Bialer, Serwyn and Mandelbaum, Michael, *The Global Rivals: The Soviet–American Contest for Supremacy* (London, I. B. Tauris, 1989).

Bille, Matt and Lishock, Erika, *The First Space Race* (College Station, TX, A&M University Press, 2004).

Bond, Peter, *The Continuing Story of the International Space Station* (Chichester, Springer Praxis, 2002).

Bowen, Wyn, 'Missile defence and the transatlantic security relationship', *International Affairs*, Vol. 77, No. 3 (July, 2001).

Brooks, H., 'The motivations for space activity', in A. A. Needell (ed.), *The First Twenty-Five Years in Space: A Symposium* (Washington, DC, Smithsonian Institution Press, 1983).

Brown, D. A., 'India developing INSAT follow-on to serve 1990's national needs', *Aviation Week and Space Technology* (12 May 1986).

Bulkely, Rip and Spinardi, Graham, *Space Weapons: Deterrence or Delusion?* (Cambridge, Polity Press, 1986).

Burchill, Scott, 'Liberalism', in Scott Burchill, Andrew Linklater, Richard Devetak, Jack Donnelly, Matthew Paterson, Christian Reus-Smit and Jacqui True, *Theories of International Relations* (3rd edn) (London, Palgrave-Macmillan, 2005), pp. 55–83.

Burrows, William, *Deep Black* (London, Transworld Publishers, 1987).

Buzan, Barry, Waever, Ole and de Wilde, Jaap, *Security: A New Framework for Analysis* (Boulder, CO, Lynne Rienner, 1998).

Byrnes, Mark A., *Politics and Space: Image Making by NASA* (Westport, CT, Praeger, 1994).

Callahan, David and Greenstein, Fred I., 'The reluctant racer: Eisenhower and US space policy', in Roger D. Launius and Howard E. McCurdy (eds), *Spaceflight and the Myth of Presidential Leadership* (Champaign, IL, University of Illinois Press, 1997).

Cash, Don E., *The Politics of Space Cooperation* (West Lafayette, IN, Purdue University Press, 1967).

Cashin, James P. and Spencer, Jeffrey D., *Space and Air Force: Rhetoric or Reality?* (Maxwell AFB, AL, Air Command and Staff College, 1999).

Caton, Jeffrey L., 'Joint warfare and military dependence on space', *Joint Forces Quarterly* (Winter, 1995–6), pp. 49–53.

Chen Lan, 'Hard road to commercial space', *Spaceflight*, Vol. 48, No. 11 (November, 2006), pp. 427–31.

Chen, Yanping, 'China's space policy – a historical review', *Space Policy*, Vol. 7 (1991), pp. 116–28.

Cheng, Bin, 'The legal regime of air space and outer space: the boundary problem', *Annals of Air and Space Law*, Vol. 5 (1980).

Chinanews.cn, 'China's space programme aims at peaceful use of space resources' (15 October 2005), available at www.chinanews.cn/news/2005/2005–10–15/12428. html.

Chipman, R., 'Multilateral intergovernmental co-operation', in *The World in Space* (Englewood Cliffs, NJ, Prentice Hall, 1982)

Chipman, R, *The World in Space* (Englewood Cliffs, NJ, Prentice Hall, 1982).

Clark, P. S., 'The Chinese space programme', *Journal of the British Interplanetary Society*, Vol. 37 (1984), p. 195

Clegg, Elizabeth and Sheehan, Michael, 'Space as an engine of development: India's space programme', *Contemporary South Asia*, Vol. 3 (1994), pp. 25–35.

CNN.com, 'Bush offers China space cooperation' (21 April 2006), available at http://edition.cnn.com/2006/TECH/space/04/21/China.us.space.reut/index.html.

Collino, R., 'The US space program: an international viewpoint', *International Security*, Vol. 11 (1987).

Committee on Commerce, Science and Transportation, United States Senate, *Soviet Space Programs 1976–80, Part I* (Washington, DC, US Government Printing Office, 1982).

Committee on Space Policy, National Academy of Sciences/National Academy of Engineering, 'Towards a new era in space: realigning US policies to new realities', *Space Policy*, Vol. 5, No. 3 (August 1989).

Cooper, Robert S., 'No sanctuary: a defense perspective on space', *Issues in Science and Technology*, Vol. 2, No. 3 (Spring, 1986). Reprinted in US Department of Defense, *Earlybird: Special Edition* (Washington, DC, US Department of Defense, 17 June 1986).

Correll, John T., 'Destiny in space', *Air Force Magazine* (August, 1998).

Cox, Robert, 'Social forces, states and world orders: beyond international relations theory', in Robert O. Keohane (ed.), *Neorealism and Its Critics* (New York, Columbia University Press, 1986), pp. 204–54.

Crawley, E. and Rymarcsuk, J., 'US–Soviet cooperation in space – benefits, obstacles and opportunities', *Space Policy*, Vol. 8 (1992), pp. 29–38.

Cremins, Thomas E., 'Security in the space age', *Space Policy*, Vol. 6 (1990), pp. 33–44.

Curien, Hubert, 'For peace or for war? Competition in the control of outer space', *NATO's Fifteen Nations*, Vol. 27 (April–May, 1982), pp. 18–20.

Dallek, Robert, *John F. Kennedy: An Unfinished Life* (London, Penguin, 2004).

Daniloff, Nicholas, *The Kremlin and the Cosmos* (New York, Alfred A. Knopf, 1972).

Davenport, Richard P., *Strategies for Space: Past, Present and Future* (Newport, RI, Naval War College, 1988).

Day, Dwayne A., 'Cover stories and hidden agendas: early American space and national security policy', in Roger D. Launius, John M. Logsdon and Robert W. Smith (eds), *Reconsidering Sputnik: Forty Years Since the Soviet Satellite* (London, Routledge, 2000).

Day, Dwayne A. and Burgess, Colin, 'Monkey in a blue suit', *Spaceflight*, Vol. 48, No. 7 (July, 2006).

DeBlois, Bruce M., 'Space sanctuary: a viable national strategy', *Airpower Journal*, Vol. 12, No. 4 (Winter, 1998), pp. 41–7.

DeBlois, Bruce M., Garwin, Richard L., Kemp, R. Scott and Marwell, Jeremy C., 'Space weapons: crossing the US Rubicon', *International Security*, Vol. 29, No. 2 (Fall, 2004), pp. 50–84.

Defense Intelligence Agency, *Soviet Military Space Doctrine* (Washington DC, Defense Intelligence Agency, 1984).

Dellios, Rosita, 'China's space programme: a strategic and political analysis', *Culture Mandala*, Vol. 7, No. 1 (December, 2005), available at www.international-relations.com/CM7–1Wb/ChinasSpaceWB.htm.

Demac, D., *Tracing New Orbits* (New York, Columbia University Press, 1986).

Dembling, Paul G., 'Commercial utilisation of space and the law', *Yearbook of Air and Space Law* (1967), pp. 283–95.

Department of Defense, Joint Publication 1, *Joint Warfare of the Armed Forces of the United States* (Washington, DC, Department of Defense, 1995).

Department of the Air Force, AFM 1-1, *Basic Aerospace Doctrine of the United States Air Force, 1992* (Maxwell AFB, AL, Department of the Air Force, 1992).

Dickson, Anna K., *Development and International Relations: A Critical Introduction* (Cambridge, Polity Press, 1997).

Din, Allan M., 'Stopping the arms race in outer space', *Journal of Peace Research*, Vol. 20 (1983), pp. 221–5.

Divine, Robert A., *The Johnson Years, Vol. 2* (Lawrence, KS, University Press of Kansas, 1987).

Dockrill, Saki, *Eisenhower's New Look National Security Policy, 1953–61* (London, Macmillan Press, 1996).

Dolman, Everett C., *Astropolitik: Classical Geopolitics in the Space Age* (London, Frank Cass, 2002).

Donnelly, Jack, 'Realism', in Scott Burchill, Andrew Linklater, Richard Devetak, Jack Donnelly, Matthew Paterson, Christian Reus-Smit and Jacqui True, *Theories of International Relations* (3rd edn) (London, Palgrave-Macmillan, 2005), pp. 29–54.

Dougherty J. and Pfaltzgraff, R., *Contending Theories of International Relations* (New York, Harper & Rowe, 1981).

Dougherty, Vice-Admiral William, 'Storm from space', *Proceedings of the US Naval Institute* (August, 1992).

Downing, J., 'Cooperation and competition in satellite communication: the Soviet Union', in D. Demac (ed.), *Tracing New Orbits – Cooperation and Competition in Global Satellite Development* (New York, Columbia University Press, 1986).

Doyle, S.E., 'Legal practicalities and realities of the use of outer space by developing countries', *Proceedings of the 32nd Colloquium of the Law of Outer Space* (1989).

Dupas, A., 'Asia in space: the awakening of China and Japan', *Space Policy*, Vol. 4 (1988), pp. 31–40.

Durch, J.W. and Wilkening, D. A., 'Steps into space', in J. W. Durch (ed.), *National Interests and the Military Use of Space* (Cambridge, MA, Ballinger Publishing Co., 1984).

Economic Commission for Africa, *African Information Society Initiative* (1996).

Elkin, J. and Fredericks, B., 'Military implications of India's space programme', *Air University Review*, Vol. 34 (1983), pp. 56–63.

Emme, E. M. (ed.), *The History of Rocket Technology* (Detroit, MI, Wayne State University Press, 1964).

Estes III, Howell M., 'Space and joint space doctrine', *Joint Force Quarterly* (Winter, 1996–7).

Etzioni, A., 'Comments', in A. A. Needell (ed.), *The First Twenty-Five Years in Space: A Symposium* (Washington, DC, Smithsonian Institution Press, 1983).

European Commission, *White Paper (COM[2003]673)*, 'Space: a new frontier for an expanding union' (Luxembourg, European Communities, 2003).

European Space Agency, *Green Paper on European Space Policy: Report on the Consultation Process* (BR-208) (Noordwijk, ESA Publications Division, 2003).

Executive Office of the President, *A National Security Strategy for a New Century* (Washington, DC, The White House, December, 1999).

Fairley, Peter, *Man on the Moon* (London, Mayflower Books, 1969).

Farber, David (ed.), *The Sixties: From Memory to History* (London, University of North Carolina Press, 1994).

Fierke, K. M., 'Constructivism', in Tim Dunne, Milja Kurki and Steve Smith (eds), *International Relations Theories: Discipline and Diversity* (Oxford, Oxford University Press, 2007).

Fischer, P., 'The origins of the Federal Republic of Germany's space policy 1959–1965 – European and National Dimensions', *ESA HSR-12* (Noordwijk, ESA, January, 1994).

Flavell, Paula, 'Tenets of air and space power: a space perspective', *Air and Space Power Journal* (Summer, 2004), p. 1, web offprint.

Foreign Relations of the United States 1955–57, Vol. IV, Western European Security and Integration (Washington, DC, US Government Printing Office, 1986).

Franke III, Henry G., *An Evolving Joint Space Campaign Concept and the Army's Role* (Fort Leavenworth, KS, Army Command and General Staff College, 1992).

Frutkin, Arnold, 'International cooperation in space', *Science* (24 July 1970).

Frye, A., 'Soviet space activities: a decade of pyrric politics', in L. P. Bloomfield (ed.), *Outer Space: Prospects for Man and Society* (New York, Frederick Praeger, 1968).

Furniss, Tim, *One Small Step: The Apollo Missions, The Astronauts, the Aftermath – A Twenty Year Perspective* (Yeovil, Haynes Publishing Group, 1989).

Garver, John W., 'China's response to the SDI', *Asian Survey* (November, 1986), pp. 1220–39.

Garwin, Richard, 'Space defense: the impossible dream?', *NATO's Sixteen Nations* (April, 1986), pp. 22–6.

Gavin, James M., *War and Peace in the Space Age* (New York, Harpers, 1958).

Geens, P., 'The new European space agency', *ESA Bulletin*, No. 4 (February, 1976).

George, Hubert, 'Developing countries and remote sensing: how intergovernmental factors impede progress', *Space Policy*, Vol. 16 (2000), pp. 267–73.

Gilks, Anne, 'China's space policy: review and prospects', *Space Policy*, Vol. 13 (1997), pp. 215–27.

Gilmartin, Patricia, 'Gulf War rekindles US debate on protecting space systems data', *Aviation Week and Space Technology*, Vol. 134, No. 17 (29 April 1991), p. 55.

Glaser, Bonnie S. and Garrett, Banning N., 'Chinese perspectives on the SDI', *Problems of Communism* (March–April 1986), pp. 28–42.

Glaser, Charles, 'Realists as optimists: cooperation as self-help', *International Security*, Vol. 19 (1994–5), pp. 50–90.

Goldsen, Joseph M. (ed.), *Outer Space in World Politics* (London, Pall Mall Press, 1963).

Golovine, M. N., *Conflict in Space: A Pattern of War in a New Dimension* (London, Temple Press Ltd, 1962).

Gorin, Peter. A., 'Rising from the cradle: Soviet perceptions of space before Sputnik', in Roger D. Launius, John M. Logsdon and Robert W. Smith (eds), *Reconsidering Sputnik: Forty Years Since the Soviet Satellite* (London, Routledge, 2000).

Gosh, S. K., 'India's space programme and its military implications', *Asian Defence Journal* (September, 1981).

Grahame, David, 'A question of intent: missile defence and the weaponisation of space', *Basic Notes* (1 May 2002), available at www.basic.org/pubs/Notes/2002NMDspace.htm.

Gray, Colin S., 'The influence of space power upon history', *Comparative Strategy* (October–December, 1996), pp. 293–308.

Grey, J., 'The international team', in J. Grey, *Beachheads in Space* (New York, Macmillan, 1983).

Groen, Bram and Hampden-Turner, Charles, *The Titans of Saturn* (Singapore, Marshall Cavendish, 2005).

Hall, R. Cargill, and Neufeld, Jacob (eds), *The United States Air Force in Space, 1945 to the 21st Century* (Andrews AFB, MD, USAF History and Museums Programme, 1995).

Hanberg, Roger, 'Rationales of the space program', in Eliger Sadeh (ed.), *Space Politics and Policy: An Evolutionary Perspective* (New York, Kluwer Academic Publishers, 2002).

Harford, James, *Korolev: How One Man Masterminded the Soviet Drive to Beat America to the Moon* (New York, John Wiley & Sons, 1997).

Harford, James, 'Korolev's triple play: Sputniks 1, 2 and 3', in Roger D. Launius, John M. Logsdon and Robert W. Smith (eds), *Reconsidering Sputnik: Forty Years Since the Soviet Satellite* (London, Routledge, 2000).

Harris, Ray, *Satellite Remote Sensing* (London, Routledge and Kegan Paul, 1987).

Harris, Ray, 'Current policy issues in remote sensing: report by the International Policy Advisory Committee of ISPRS', *Space Policy*, Vol. 19 (2003), pp. 293–6.

Harvey, Brian, *Race into Space: The Soviet Space Programme* (Chichester, Ellis Horwood, 1988).

Harvey, Brian, *The Japanese and Indian Space Programmes: Two Roads into Space* (Chichester, Springer Praxis Books, 2000).

Harvey, Brian, 'China's space programme: emerging competitor or potential partner?', Center for Nonproliferation *Occasional Paper, No. 12* (Monterey, CA, 2003).

Harvey, David, *The Condition of Postmodernity* (Oxford, Blackwell, 1990).

Harvey, Dodd L. and Ciccoritti, Linda C., *US–Soviet Cooperation in Space* (Miami, FL, Center for Advanced International Studies, University of Florida, 1974).

Hastings, Max and Jenkins, Simon, *The Battle for the Falklands* (New York, W. W. Norton, 1983).

Hays, Peter and Mueller, Karl, 'Going boldly – where? Aerospace integration, the Space Commission, and the Air Force's vision for space', *Aerospace Power Journal* (2001), pp. 34–49.

Heppenheimer, T. A,, *Countdown: A History of Space Flight* (New York, John Wiley and Sons, 1997).

Herz, John, 'Idealist internationalism and the security dilemma', *World Politics*, Vol. 2 (1950), pp. 157–80.

Higgins, Trumbull, *The Perfect Failure: Kennedy, Eisenhower, and the CIA at the Bay of Pigs* (New York, W. W. Norton, 1987).

Hitchens, Theresa, 'Weapons in space: silver bullet or Russian roulette?', *Center for Defense Information* (Washington, DC, *Center for Defense Information*, 18 April 2002).

Hoagland, J., 'The other space powers: Europe and Japan', in U. Ra'anan and R. Pfaltzgraff (eds), *International Security Dimensions of Space* (Hamden, CT, Archon Books, 1984).

Holsti, K. J., *International Politics: A Framework for Analysis* (5th edn) (Englewood Cliffs, NJ, Prentice-Hall, 1988).

Holtzmann Kelvas, Bettyann, *Almost Heaven: The Story of Women in Space* (Cambridge, MA, MIT Press, 2005).

House of Representatives, *Conference Report on S.1059, National Defense Authorisation Act for Fiscal Year 2000*, 106th Congress, 1st session (5 August 1999), H.R. 106-301, sec. 1621-30, 'Commission to Assess United States National Security Space Management and Organisation', available at http;//www.nuclearfiles.org/kmissiledefense/intro.html.

Hufbauer, K., 'Solar observational capabilities and the solar physics community since Sputnik', in M. Collins and S. Fries, *A Spacefaring Nation* (Washington, DC, Smithsonian Institution Press, 1991).

Hui Zhang, 'Action–reaction: US space weaponisation and China', *Arms Control Today* (December, 2005), available at http://www.armscontrol.org/act/2005_12?Dec-cvr.asp?print.

Hulsroj, Peter, 'Beyond global: the international imperative of space', *Space Policy*, Vol. 18 (2002).

Humble, R. D., *The Soviet Space Programme* (London, Routledge, 1988).

Hyten, John E., 'A sea peace or a theater of war?', *Air and Space Power Journal* (Fall, 2002), pp. 78–92.

Ince, Martin, *Space* (London, Sphere Books, 1981).

Indian Space Research Organisation, *Department of Space, Annual Report, 2005–6*.

Ions, Edmund S., *The Politics of John F. Kennedy* (London, Routledge and Kegan Paul, 1967).

Isaacs, Jeremy and Downing, Taylor, *Cold War* (London, Bantam Press, 1998).

Jakhu, Ram S., 'The legal status of the geostationary orbit', *Annals of Air and Space Law*, Vol. 7 (1982).

Jane's Spaceflight Directory (London, Jane's, 1987).

Jasani, Bhupendra *Space Weapons and International Security* (Oxford, Oxford University Press, 1987).

Jasani, Bhupendra, 'US national missile defence and international security: blessing or blight?', *Space Policy*, Vol. 17 (2001), pp. 243–7.

Jasani, Bhupendra, and Lee, Christopher, *Countdown to Space War* (London, Taylor & Francis, 1984).

Jasani, Bhupendra, and Toshiba Sakata (eds), *Satellites for Arms Control and Crisis Monitoring* (Oxford, Oxford University Press, 1987).

Jasentuliyana, N., 'Review of the recent discussions relating to aspects of article one of the Outer Space Treaty', in *Proceedings of the 29th Colloquium of the Law of Outer Space* (1987).

Jasentuliyana, N., 'Ensuring equal access to the benefits of space technologies for all countries', *Space Policy*, Vol. 10 (1994), pp. 7–18.

Jayaraman, K. S., 'India's Space Agency proposes manned spaceflight program', *Space News* (10 November 2006), available at www.space.com/news/061110_india_mannedspace.htm.

Johnson, N. L., *Soviet Military Strategy in Space* (London, Jane's, 1987).

Johnson, Oris B. Major-General, 'Space: today's first line of defense', *Air University Review*, Vol. 20, No. 1 (November–December, 1968), pp. 95–102.

Johnson-Freese, Joan, 'Scorpions in a bottle: China and the US in space', *Nonproliferation Review*, Vol. 11, No. 2 (Summer, 2004), pp. 166–82.

Johnson-Freese, Joan and Erickson, Andrew S., 'The emerging China–EU space partnership: a geotechnological balancer', *Space Policy*, Vol. 22 (2006), pp. 12–22.

Jusell, Judson J., *Space Power Theory: A Rising Star* (Maxwell AFB, AL, Air Command and Staff College, 1998).

Karash, Y., *The Superpower Odyssey: A Russian Perspective on Space Cooperation* (Reston, VA, American Institute of Aeronautics, and Astronautics, 1999).

Karp, Aaron, 'Ballistic missiles in the Third World', *International Security*, Vol. 9 (1984–5), pp. 166–95.

Kaul, K., 'India in space', *Strategic Analysis* (New Delhi), Vol. 10 (1986), pp. 853–63.

Kay, W. D., *Defining NASA: The Historical Debate over the Agency's Mission* (Albany, NY, University of New York Press, 2006).

Keaney, Thomas A., and Cohen, Elliot A., *The Gulf War Air Power Summary Report* (Washington, DC, US Government Printing Office, 1993).

Kennedy, President John F., *Text of address to Congress entitled, 'Special Message to the Congress on Urgent National Needs'* (25 May 1961) JFK Library, available at http://www.cs.umb.edu/jfklibrary/j052561.htm.

Kennedy, President John F., *Address Before the 18th General Assembly of the United Nations*, New York (20 September 1963), JFK Library, available at www.cs.umb.edu/jfklibrary/j092063.htm.

Killian, J. R., *Sputnik, Scientists and Eisenhower* (Cambridge, MA, MIT Press, 1977).

Kingwell, Jeff, 'The militarization of space: a policy out of step with world events', *Space Policy*, Vol. 6 (1990), pp. 107–11.

Kingwell, Jeff, 'The rise of the Asian space dragon', *Space Policy*, Vol. 10 (1994), pp. 185–6.

Kirby, Stephen, and Robson, Gordon (eds), *The Militarisation of Space* (Brighton, Wheatsheaf Books, 1987).

Knorr, Klaus, 'The international implications of outer space activities', in Joseph M. Goldsen (ed.), *Outer Space in World Politics* (London, Pall Mall Press, 1963).

Kolovos, A, 'Why Europe needs space as part of its security and defence policy', *Space Policy*, Vol. 18 (2002), pp. 257–61.

Kraft, Chris, *Flight: My Life in Mission Control* (New York, Plume/Penguin Books, 2002).

Kranz, Gene, *Failure is Not an Option: Mission Control from Mercury to Apollo 13 and Beyond* (New York, Berkley Books, 2001).

Krige, J., 'The prehistory of ESRO 1959/60', *ESA HSR-1* (Noordwijk, ESA, July, 1992).

Krige, J., 'The launch of ELDO', *ESA HSR-7* (Noordwijk, ESA, 1993).

Krige, J., 'Europe into space: the Auger years 1959–1967', *ESA HSR-8* (Noordwijk, ESA, 1993).

Lambakis, Steven, 'Space control in Desert Storm and beyond', *Orbis* (Summer, 1995).

Lapp, R. E., *Man and Space* (London, Secker & Warburg, 1961).

Launius, Roger D., 'Historical dimensions of the space age', in Eligar Sadeh (ed.), *Space Politics and Policy: An Evolutionary Perspective* (New York, Kluwer Academic Publishers, 2002).

Launius, Roger D., Logsdon, John M. and Smith, Robert W. (eds), *Reconsidering Sputnik: Forty Years Since the Soviet Satellite* (London, Routledge, 2000).

Laver, Michael, 'The politics of inner space: tragedies of three commons', *European Journal of Political Research*, Vol. 12 (1984).

Layton, C., 'The European space effort in the light of global European policy', *ESA Bulletin*, No. 4 (1976).

Lebow, Richard Ned, 'Classical realism', in Tim Dunne, Milja Kurki and Steve Smith (eds), *International Relations Theories: Discipline and Diversity* (Oxford, Oxford University Press, 2007).

Lee, Christopher, *War in Space* (London, Hamish Hamilton, 1986).

Lee, James G., *Counterspace Operations for Information Dominance* (Maxwell AFB, AL, Air University Press, 1994).

Lee, Min, 'China aims to put man on the moon by 2020', Space.com (27 November 2005), available at www.space.com/missionlaunches/ap_051127_china_moon.html.

Leskov, S., 'Soviet space in transit', *Space Policy* (August, 1989).

Levine, Alan J., *The Missile and Space Race* (Westport, CT and London, Praeger, 1994).

Lewis, James A., 'China as a military space competitor', Center for Strategic and International Studies (August, 2004).

Lewis, Jeffrey, 'Engage China, engage the world', *adAstra* (May, 2005), available at www.space.com/adastra/china_engagement_0505.html.

Lewis, John Wilson and Xue Litai, *China's Strategic Seapower: The Politics of Force Modernisation in the Nuclear Age* (Stanford, CA, Stanford University Press, 1994).

Lewis, S. J. and Lewis, R.A, *Space Resources: Breaking the Bonds of Earth* (New York, Columbia University Press, 1987).

Lider, Julian, *Correlation of Forces* (Aldershot, Gower, 1986).

Logsdon, J., 'Introduction', in A. A. Needell (ed.), *The First Twenty-five Years in Space: A Symposium* (Washington, DC, Smithsonian Institution Press, 1983).

Logsdon, J., 'Evaluating Apollo', *Space Policy*, Vol. 5, No. 3 (August, 1989), pp. 188–92, 36–43.

Logsdon J. and Dupas, A., 'Was the race to the moon real?', *Scientific American* (June, 1994).

Long Range Plan: Implementing USSPACECOM Vision for 2020 (Peterson AFB, CO, US Space Command, March, 1998).

Longden, Norman and Guyenne, Duc (eds), *Twenty Years of European Cooperation in Space: An ESA Report* (Noordwijk, ESA Scientific and Technical Publications Branch, 1984).

Lord, General Lance, Commander, USAF Space Command, 'The argument for space superiority: remarks prepared for Air War College' (Maxwell AFB, AL, 29 March 2004), available at www.peterson.af.mil/hqafspc/Library/speeches/Speeches.asp? YearList=2004.

Lust, R., 'European cooperation in space', *ESA Bulletin*, No. 58 (Noordwijk, ESA, May, 1989).

McCurdy, Howard, 'The decision to build the space station: too weak a commitment?', *Space Policy*, Vol. 4, No. 4 (November, 1988).

McCurdy, Howard, *The Space Station Decision: Incremental Politics and Technological Choice* (Baltimore, MD, Johns Hopkins University Press, 1990).

McDougall, Walter A., 'Scramble for space', *Wilson Quarterly*, Vol. 4 (Autumn, 1980), pp. 71–82.

McDougall, Walter A., 'Sputnik, the space race and the Cold War', *Bulletin of the Atomic Scientists*, Vol. 41, No. 5 (1985).

McDougall, Walter A., *The Heavens and the Earth: A Political History of the Space Age* (New York, Basic Books, 1985).

McKinley, Cynthia A. S., 'When the enemy has our eyes', available at http://fas.org/spp/eprint/mckinley.htm.

Mackintosh, J. M., *Strategy and Tactics of Soviet Foreign Policy* (Oxford, Oxford University Press, 1962).

Madders, Kevin, *A New Force at a New Frontier* (Cambridge, Cambridge University Press, 1997).

McLean, Alasdair, *Western European Military Space Policy* (Aldershot, Dartmouth Publishing, 1992).

McLean, Alasdair, 'A new era? Military space policy enters the mainstream', *Space Policy*, Vol. 16 (2000), pp. 243–7.

McNiel, Samuel, 'Proposed tenets of space power: six enduring truths', *Air and Space Power Journal* (Summer, 2004).

Madders, K. and Thiebaut, W., 'Two Europes in one space: the evolution of relations between the European Space Agency and the European Community in space affairs', *Journal of Space Law*, Vol. 20, No. 2 (1992), pp. 117–32.

Mahnken, Thomas, 'Why Third World space systems matter', *Orbis*, Vol. 35 (1991), pp. 563–79.

Malik, J. Mohan, 'India copes with Kremlin's fall', *Orbis*, Vol. 37 (1993), pp. 69–87.

Mama, H. P., 'India's space programme: across the board on a shoestring', *Interavia*, Vol. 35 (January, 1980), pp. 60–2.

Manno, J., *Arming the Heavens: The Hidden Military Agenda for Space, 1945–1995* (New York, Dodd, Mead and Co., 1984).

Mantz, Michael R., *The New Sword: A Theory of Space Combat Power* (Maxwell AFB, AL, Air University Press, 1995).

Mark, Hans, *The Space Station: A Personal Journey* (Durham, NC, Duke University Press, 1987).

Markoff, John, 'Outer space: the military's new high ground', *Baltimore Sun* (6 July 1980).

Martel, William C. and Toshi Yoshihara, 'Averting a Sino–US space race', *Washington Quarterly*, Vol. 26, No. 4 (Autumn, 2003), pp. 19–35.

Martin, Lisa A., 'Neoliberalism', in Tim Dunne, Milja Kurki and Steve Smith (eds), *International Relations Theories: Discipline and Diversity* (Oxford, Oxford University Press, 2007).

Marwah, Onkar, 'India's nuclear and space programs: intent and policy', *International Security* (Fall, 1977).

Mearsheimer, John J., 'Structural realism', in T. Dunne, M. Kurki, and S. Smith (eds), *International Relations Theories* (Oxford, Oxford University Press, 2007), pp. 71–88.

Memorandum of Understanding Between the National Aeronautics and Space Administration of the United States of America and the Canadian Space Agency Concerning Cooperation on the Civil International Space Station. Signed 29 January 1998, Washington, DC, available at ftp://ftp.hq.nasa.gov/pao/reports/1998/nasa_csa.html.

Memorandum of Understanding Between the National Aeronautics and Space Administration of the United States of America and the European Space Agency Concerning Cooperation on the Civil International Space Station. Signed 29 January 1998, Washington, DC, available at ftp://ftp.hq.nasa.gov/pao/reports/1998/nasa_esa.html.

Memorandum of Understanding Between the National Aeronautics and Space Administration of the United States of America and the Government of Japan Concerning Cooperation on the Civil International Space Station. Signed 24 February 1998, Washington, DC, available at ftp://ftp.hq.nasa.gov/pao/reports/1998/nasa_Japan.html.

Memorandum of Understanding Between the National Aeronautics and Space Administration of the United States of America and the Russian Space Agency Concerning Cooperation on the Civil International Space Station. Signed 29 January 1998, Washington, DC, available at ftp://ftp.hq.nasa.gov/pao/reports/1998/nasa_Russian.html.

Mikheyev, D., *The Soviet Perspective on the Strategic Defense Initiative* (Washington, DC, Pergamon-Brassey's, 1987).

Miles, E., 'Transnationalism in space: inner and outer', *International Organisation*, Vol. 25 (1971), pp. 602–25.

Milholin, G., *Research Report: West German Aid to India's Rocket Programme* (Washington, DC, Wisconsin Project on Nuclear Arms Control, April, 1989).

Mingst, Karen, 'Functionalist and regime perspectives: the case of Rhine river cooperation', *Journal of Common Market Studies*, Vol. 20 (1981).

Mitrany, David, 'The functional approach in historical perspective', *International Affairs*, Vol. 47 (1971), pp. 532–43.

Mohan, C. Raju and Subrahmanyam, K., 'High technology weapons in the developing world', in Eric H. Arnott (ed.), *New Technologies for Security and*

Arms Control: Threats and Promises (Washington, DC, American Association for the Advancement of Science, 1989).

Moore, George M., Budura, Vic and Johnson-Freese, Joan, 'Joint space doctrine: catapulting into the future', *Joint Forces Quarterly* (Summer, 1994).

Morris, Willaim R., 'The role of China's space program in its national development strategy', *Maxwell Paper No. 24* (Maxwell AFB, AL, Air War College, August, 2001).

Mrazek, Robert J., 'Rethinking national and global security – the role of space-based observations', *Space Policy*, Vol. 5 (1989), pp. 155–63.

Murray, Bruce, 'Can space exploration survive the end of the Cold War?', *Space Policy*, Vol. 7 (1991), pp. 23–34.

Murray, Charles and Cox, Catherine, *Apollo: The Race to the Moon* (London, Secker & Warberg, 1989).

National Aeronautics and Space Act, 85th Congress (29 July 1958).

National Security Council, *Preliminary US Policy on Outer Space*, NSC5814/1, 18 August 1958.

Navias, M., 'Ballistic missile proliferation in the Third World', *Adelphi Paper 252* (London, International Institute for Strategic Studies, 1990).

Newman, R. T., 'Propaganda: an instrument of foreign policy', *Columbia Journal of International Affairs*, Vol. 5 (1951).

Nolan, Janne E., *Trappings of Power: Ballistic Missiles in the Third World* (Washington, DC, Brookings Institution, 1991).

Nugent, Nicholas, 'The defence preparedness of India: arming for tomorrow', *Military Technology* (March, 1991), pp. 27–36.

Oberg, James E., *Red Star in Orbit* (New York, Random House, 1981).

Oberg, James E., *The New Race for Space* (Harrisburg, PA, Stackpole Books, 1984).

Oberg, James E., 'Russia's space program running on empty' (1995), available at http://astronautix.com/articles/ruspart2.htm.

Oberg, James E., *Space Power Theory* (Colorado Springs, CO, US Air Force Academy, 1999).

Office of the US Secretary of Defense, Annual Report to Congress: The Military Power of the People's Republic of China, available at http://www.globalsecurity.org/military/library/report/2004/d20040528prc.pdf.

O'Neill, William L., *Coming Apart: An Informal History of America in the 1960's* (New York, Time Books/Random House, 1971).

Onuf, Nicholas, *World of Our Making: Rules and Rule in Social Theory and International Relations* (Columbia, SC, University of South Carolina Press, 1989).

Osman, T., *Space History* (London, Michael Joseph, 1983).

OTA, *International Cooperation and Competition in Civilian Space Activities* (Washington, DC, US Congress, Office of Technology Assessment, OTA-ISC-239, July, 1985).

Parkinson, Robert, *Citizens of the Sky* (Stotfold, 2100 Publishing, 1987).

Peebles, Curtis, *Battle for Space* (London, Book Club Associates, 1983.

Peebles, Curtis, *The High Frontier; The US Air Force and Military Space Programme* (Washington, DC, Air Force History and Museums Program, 1997).

Pelton, Joseph N., 'The proliferation of communications satellites: gold rush in the Clarke Orbit', in James E. Katz (ed.), *People in Space* (Oxford, Transaction Books, 1985).

Peoples, Columba, 'Haunted dreams: critical theory and the militarisation of space', Paper presented at the Annual Conference, British International Studies Association (Cork, December, 2006).

Perry, G., 'Perestroika and glasnost in the Soviet space programme', *Space Policy* (November, 1989).

Perry, Geoffrey E., 'Russian hunter–killer satellite experiments', *Military Review* (October, 1978).

Peterson, M. J., *International Regimes for the Final Frontier* (New York, State University of New York Press, 2005).

Pike, J., 'US and Soviet BMD programmes', in B. Jasani (ed.), *Outer Space: A Source of Conflict or Cooperation?* (Tokyo, United Nations University Press, 1991), p. 185.

Pollack, Herman, 'International relations in space: a US view', *Space Policy*, Vol. 4 (1988), pp. 24–30.

Popescu, J., *Russian Space Exploration: The First 21 Years* (Oxford, Gothard House, 1979).

Quistgaard, E., 'ESA and Europe's future in space', *ESA Bulletin*, No. 39 (August, 1984).

Radio Regulations Article S23.13, and Appendix 30.

Rajan, Mohan Sundara, *India in Space* (New Delhi, Government of India, Ministry of Information and Broadcasting, 1979).

Rajan, Y. S., 'Benefits from space technology: a view from a developing country', *Space Policy*, Vol. 4 (1988).

Ramo, S., 'The practical dimensions of space', in A. A. Needell (ed.), *The First Twenty-five Years in Space: A Symposium* (Washington, DC, Smithsonian Institution Press, 1983).

RAND Corporation, *Preliminary Design for an Experimental World-Circling Spaceship* (Santa Monica, CA, Project RAND, 1946, Reprinted RAND Corporation, 1999).

Rao, U. R., 'Space technology in developing nations: an assessment', *Space Policy* (1993).

Reagan Administration Space Policy Statement, White House (Washington, DC, 4 July, 1982). Reproduced as Appendix 6 in Hans Mark, *The Space Station: A Personal Journey* (Durham, NC, Duke University Press, 1987), pp. 243–7.

Report of the Commission to Assess United States National Security Space Management and Organisation (Washington, DC, USGPO, 11 January 2001), available at www.space.gov/docs/fullreport.pdf.

Robb, Charles S., 'Star wars II', *Washington Quarterly*, 221 (Winter, 1999), pp. 81–6.

Robertson, A. H., *European Institutions* (London, Stevens & Sons, 1973).

Roederer, Juan G. 'The participation of developing countries in space research', *Space Policy*, Vol. 1 (1985)

Rosas, Allan, 'The militarisation of space and international law', *Journal of Peace Research*, Vol. 20 (1983), pp. 357–64.

Royal Institute of International Affairs, *Europe's Future in Space* (London, Routledge and Kegan Paul, 1988).

'Russia and China's Moon Pact', *Spaceflight*, Vol. 48, No. 11 (November, 2006), p. 405.

Sadeh, Eligar, Lester, James P. and Sadeh, Willy Z., 'Modelling international cooperation for space exploration', *Space Policy*, Vol. 12 (1996), pp. 207–23.

Saunders, Phillip C., 'China's future in space: implications for US security', *adAstra: The Magazine of the National Space Society* (20 September, 2005), available at http://www.space.com/adastra/China_implications_0505.html.

Schaeur, William H., *The Politics of Space: A Comparison of the Soviet and American Space Programmes* (New York, Holmes & Meier, 1976).

Scheffren, Jurgen, 'Peaceful and sustainable use of space – principles and criteria for evaluation', in Wolfgang Bender, Regina Hagen, Martin Kalinowski and Jurgen Scheffren (eds), *Space Use and Ethics*, Vol. I (Münster, Agenda-Verlag, 2001).

Schefter, James, *The Race: The Complete Story of How America Beat Russia to the Moon* (New York, Anchor Books, 2000).

Schull, Todd C., 'Space operations doctrine: the way ahead', *Air and Space Power Journal* (Summer, 2004).

Schwartz, M., 'The politics of European space collaboration', *Intermedia*, Vol. 9, No. 4 (July, 1981).

Senate Resolution 327, Report No. 1925, 85th Congress, 2nd session (24 July 1958).

Serafimov, K. B., 'Achieving world-wide co-operation in space', *Space Policy*, Vol. 5 (1989).

Sevastyanov, V. and Ursol, A., 'Cosmonautics and social development', *International Affairs* (Moscow) (November, 1977), pp. 70–7.

Shayler, David J. and Moule, Ian, *Women in Space; Following Valentina* (Chichester, Springer Praxis, 2005).

Sheehan, Michael, *The Arms Race* (Oxford, Martin Robertson, 1983).

Sheehan, Michael, *Arms Control: Theory and Practice* (Oxford, Basil Blackwell, 1988).

Sheehan, Michael, 'Chinese military space policy', *Enjeux Atlantiques*, Vol. 4 (1991), pp. 43–5.

Sheehan, Michael, *Balance of Power: History and Theory* (London, Routledge, 1996).

Sheehan, Michael, 'Mars calling: China's space programme', *The World Today* (November, 2003), pp. 16–17.

Sheehan, Michael, *International Security: An Analytical Survey* (Boulder, CO, Lynne Rienner, 2005).

Siddiqi, Asif A., 'Korolev, Sputnik and the International Geophysical Year', in Roger D. Launius, John M. Logsdon and Robert W. Smith (eds), *Reconsidering Sputnik: Forty Years Since the Soviet Satellite* (London, Routledge, 2000).

Siddiqi, Asif A., *Sputnik and the Soviet Space Challenge* (Gainesville, FL, University of Florida Press, 2003).

Simpson, Theodore R. (ed.), *The Space Station: An Idea Whose Time Has Come* (New York, IEEE Press, 1985).

Singh, Jasjit, 'Indian security: a framework for national strategy', *Strategic Analysis* (New Delhi, November, 1987), pp. 183–94.

Skuridin, G. A., *Entrance of Mankind into Space* (Washington, DC, National Aeronautics and Space Administration, 1976).

Smith, Andrew, *Moondust* (London, Bloomsbury, 2005).

Smith, M. V., 'Ten propositions regarding spacepower', *Fairchild Paper* (Maxwell AFB, AL, Air University Press, 2002).

Smith, M. L., 'Equitable access to the orbit/spectrum resource', in *Proceedings of the 30th Colloquium of the Law of Outer Space* (Reston, VA, American Institute for Aeronautics and Astronautics, 1988).

Smith, M. S., 'Evolution of the Soviet space programme from Sputnik to Salyut and beyond', in U. Ra'anan and R. L. Pfaltzgraff (eds), *International Security Dimensions of Space* (North Haven, CT, Archon Books, 1984).

Smith, Bob (Senator), 'The challenge of space power', *Aerospace Power Journal*, Vol. 13, No. 1 (Spring, 1999), pp. 32–9.

Sokolovski, V. D. (ed.), *Soviet Military Strategy* (Santa Monica, CA, RAND Corporation, 1963).

Sorenson, Theodore, *Kennedy* (London, Hodder & Stoughton, 1965).

Soroos, Marvin S., 'The commons in the sky: the radio spectrum and geosynchronous orbit as issues in global policy', *International Organisation*, Vol. 36 (1982).

Spanier, J. W., *Games Nations Play* (New York, Praeger, 1978).

Spanier, J. W., *Games Nations Play* (New York, Holt, Reinhart & Winston, 1984).

Specter, C., 'ISY: an opportunity to relate earth observation activities to the benefits and interests of the developing countries', *Proceedings of the 32nd Colloquium on the Law of Outer Space* (1990).

Spero, Joan, *The Politics of International Economic Relations* (London, Unwin Hymen, 1985).

Spires, David N., *Beyond Horizons: A Half Century of Air Force Space Leadership* (Maxwell AFB, AL, Air University Press, 1998).

Stapp, Richard S., *Space Dominance: Can the Air Force Control Space?* (Maxwell AFB, AL, USAF Air Command and Staff College, 1997).

Stares, P., *The Militarization of Space: US Policy 1945 to 1984* (Ithaca, NY, Cornell University Press, 1985).

Stares, Paul B., *Space Weapons and US Strategy: Origins and Development* (Beckenham, Croom Helm, 1985).

Sterner, Eric R., 'International competition and co-operation: civil space programmes in transition', *The Washington Quarterly*, Vol. 16 (1993), pp. 129–48.

Streland, Arnold H., *Clausewitz on Space: Developing Military Space Theory through a Comparative Analysis* (Maxwell AFB, AL, Air Command and Staff College, 1999).

Suzuki, Kazuto, *Policy Logics and Institutions of European Space Collaboration* (Aldershot, Ashgate, 2003).

Teets, Peter B., 'National security space: enabling joint warfighting', *Joint Force Quarterly* (Winter, 2002–3).

Teets, Peter B. (Under-Secretary of the Air Force), 'National security space in the twenty-first century', *Air and Space Power Journal* (Summer, 2004), pp. 34–7.

Thomas, Raju G. C., 'India's nuclear and space programmes: defence or development?', *World Politics*, Vol. 38 (1985–6), pp. 315–42.

Thompson, David J., 'China's military space programme: strategic threat, regional power, or national defense?', *Maxwell Paper No. 24* (Maxwell AFB, AL, Air War College, August, 2001), p. 21.

Toffler, Alvin and Toffler, Heidi, *War and Anti-War: Survival at the Dawn of the 21st Century* (New York, Little Brown & Co., 1993).

Torgerson, Thomas A., *Global Power Through Tactical Flexibility: Rapid Deployable Space Units* (Maxwell AFB, AL, Air University Press, 1994).

Trux, T., *The Space Race* (London, New English Library, 1987).

United States Air Force Doctrine Document, AFDD 2-2.1, *Counterspace Operations* (Colorado Springs, CO, USAF Doctrine Center, August, 2004).

United States Civilian Space Programmes 1958–1978, Report prepared for the Subcommittee on Space Science and Applications, Committee on Science and Technology, House of Representatives, 97th Congress, first session, Volume I (Washington, DC, US Government Printing Office, 1981).

United States Congress, Senate, Committee on Aeronautical and Space Sciences, *Documents on International Aspects of the Exploration and Use of Outer Space, 1954–1962*, Staff report, S.Doc, no. 18, 88th Congress, 1st session (1963).

United States Department of Defense, *News Transcript*, 'Secretary Rumsfeld Outlines Space Initiatives' (8 May 2001).

United States General Accounting Office, Report to the Secretary of Defense, *GAO-02-738, Military Space Operations: Planning, Funding, and Acquisition Challenges Facing Efforts to Strengthen Space Control* (Washington, DC, US Government Printing Office, September, 2002).

United States President, *Public Papers of the Presidents of the United States, John F Kennedy*, 1 January–31 December 1962 (Washington, DC, US Government Printing Office, 1963).

United States Space Command, *Long-Range Plan: Implementing USSPACECOM Vision for 2020* (Peterson AFB, CO, US Space Command, 1998), p. 7.

USAF Scientific Technology Advisory Board, 'Space technology volume', *New World Vistas: Air and Space Power for the 21st Century*, Report to the USAF Chief of Staff (Washington, DC, US Government Printing Office, 1995).

van Dyke, Vernon, *Pride and Power: The Rationale of the Space Program* (London, Pall Mall Press, 1965)

van Welck, Steven, 'India's space policy: a developing country in the space club', *Space Policy*, Vol. 3 (1987), pp. 326–34.

von Benke, Matthew J., *The Politics of Space: A History of US-Soviet Competition and Cooperation* (Boulder, CO, Westview, 1997).

von Kries, W., 'The demise of the ABM Treaty and the militarization of outer space', *Space Policy*, Vol. 18 (2002), pp. 175–8.

Walker, M., *The Waking Giant: The Soviet Union Under Gorbachev* (London, Michael Joseph, 1986).

Weber, Steve, 'Realism, détente and nuclear weapons', *International Organisation*, Vol. 44 (1990), pp. 55–82.

Weiss, Kenneth G., 'Space dragon: long march, proliferation and sanctions', *Comparative Strategy*, Vol. 18 (1999), pp. 335–60.

Wells, D. R. and Hastings, D. E., 'The US and Japanese space programmes: a comparative study', *Space Policy*, Vol. 7 (1991), pp. 233–56.

WEU Assembly, *Proceedings*, 13th Ordinary session (1966), Vol. 2 (Documents), Doc. 389.

White House, The, National Science and Technology Council, *Fact Sheet: National Space Policy* (Washington, DC, Office of the President of the United States, 19 September 1996).

Wiessner, Seigfried, 'Access to a *Res Publica Internationalis*', in *Proceedings of the 29th Coloquium of the Law of Space* (1987).

Wilford, John Noble, 'Riding high', *The Wilson Quarterly* (Autumn, 1980).

Wilhelm, Donald, *Global Communications and Political Power* (New Brunswick, NJ and London, Transaction Publishers, 1990).

Williams, Sylvia Maureen, 'Direct Broadcast satellites and international law', *International Relations*, Vol. 8 (1985).

Working Paper on Principles Regarding International Cooperation in the Exploration and Utilisation of Outer Space for Peaceful Purposes. UN Doc. A/AC.105/C.2/L.182/Rev.1 (New York, 31 March 1993).

Yong, Qiu, 'Analysis on the ASAT capability of the GMD interceptor', Presentation at the 16th International Summer Symposium on Science and World Affairs, Beijing, 17–25 July 2004. Cited in Hui Zhang, 'Action–reaction: US space weaponisation and China', *Arms Control Today* (December, 2005).

Young, H., Silcock, B. and Dunn, P., *Journey to Tranquillity* (London, Jonathan Cape, 1969).

Zabusky, Stacia E., *Launching Europe: An Ethnography of European Cooperation in Space Science* (Princeton, NJ, Princeton University Press, 1995).

Zhu Yilin and Xu Fuxiang, 'Status and prospects of china's space programme', *Space Policy*, Vol. 13 (1997), pp. 69–75.

Ziarnick, Brent D., 'The space campaign: space power theory applied to counterspace operations', *Air and Space Power Journal* (Summer, 2004), pp. 61–70.

Ziegler, David W., *Safe Havens: Military Strategy and Space Sanctuary Thought* (Maxwell AFB, AL, Air Command and Staff College, 1998).

Index